READBE FIRST for a

USER'S GUIDE TO
QUALITATIVE METHODS

Second Edition

*To our students, with thanks for all
that they have taught us about the challenges
of learning to work qualitatively.*

README FIRST for a

USER'S GUIDE TO
QUALITATIVE METHODS

Second Edition

Lyn Richards
Janice M. Morse

SAGE Publications
Thousand Oaks ▪ London ▪ New Delhi

For information:

Sage Publications, Inc.
2455 Teller Road
Thousand Oaks, California 91320
E-mail: order@sagepub.com

Sage Publications Ltd.
1 Oliver's Yard
55 City Road
London
EC1Y 1SP
United Kingdom

Sage Publications India Pvt. Ltd.
B-42, Panchsheel Enclave
Post Box 4109
New Delhi 110 017 India

Printed in the United States of America

Library of Congress Cataloging-in-Publication Data

Richards, Lyn.
Readme first for a user's guide to qualitative methods /
Lyn Richards, Janice M. Morse. — 2nd ed.
    p. cm.
Morse's name appears first on the earlier edition.
Includes bibliographical references and index.
ISBN 978-1-4129-2743-7

    1. Social sciences—Research—Methodology. 2. Qualitative research. I. Morse, Janice
M. II. Title. III. Title: Read me first for a user's guide to qualitative methods.

H62.M6612 2007
001.4'2—dc22

                                    2006024105

This book is printed on acid-free paper.

07 08 09 10 11 10 9 8 7 6 5 4 3 2

Acquisitions Editor:	Lisa Cuevas Shaw
Editorial Assistant:	Karen Greene
Production Editor:	Denise Santoyo
Copy Editor:	Cheryl Rivard
Typesetter:	C&M Digitals (P) Ltd.
Indexer:	Sylvia Coates
Cover Designer:	Bryan Fishman

Contents

Preface

Qualitative research is expanding and developing at breathtaking speed. The problems faced by us and our students a decade ago are replaced by new challenges. Where previously there were few texts available, now there are many, but most are specific to one discipline and/or one method. Where previously qualitative methods were a minority activity in most disciplines, learned as a craft in apprenticeship to experienced researchers, now they are often attempted without training, and researchers may have difficulty finding mentors. Where data handling was a gross clerical load and data access limited by human memory, computer software now provides ways of handling and analyzing data that were impossible to achieve by manual methods. The widespread use of specialist software has made qualitative research more attractive and more accessible to those without qualitative training. However, just as a decade ago, it remains difficult to begin a qualitative project, make sense of methodological choices, and get *thinking*—let alone started—on the right track. The researcher facing these difficulties is much more likely now to be facing them alone.

We wrote this book in recognition that these changes have altered the qualitative research world forever. Change continues, and we have prepared the revised edition reflecting on new pressures on researchers and new opportunities for them. We share a conviction that if the changes are merely ignored or regretted, damage is done to both the researchers and the methods. New researchers need to get over those preliminary obstacles, to understand the language of qualitative inquiry, and to know what questions to ask, where to look for information, and how to start thinking qualitatively. They also need to challenge myths and false expectations and to know what to expect before they start the "real thing." This involves placing the wonderful promises of qualitative research in the context of the methods that make them work, and the processes of choosing methods appropriate for your research. It also involves placing the promises of technology in the context of tasks and techniques.

This book is intended to be read *first* by those who are thinking about becoming qualitative researchers—before they acquire data, before they preemptively choose a method, let alone a software package, and before they commit to a project. It may be used as a text for an introductory course; or it may be used by those who are simply interested in qualitative inquiry and want to get a feel for the qualitative research process. This new edition includes a review of what software can do for you and help with finding up-to-date reviews of software and tutorials so that you may try out the computer tools and learn what they offer—and what they don't—before you propose your own project.

Above all, the aim of this book is not to teach a single method but to map the range of methods, not to commit you to one sort of research but to show you why there are so many ways of working qualitatively. As we wrote these chapters, we discovered how different our own methods were. Readers who know our work will recognize one voice or another as that of the first author of particular chapters. But all our diverse experiences pointed to the same need—a need for a book that would meet the approaching researcher at the beginning of the path into the methodological maze.

The idea for this book was developed on a postconference tour of the Krueger National Park many years ago. Forbidden from leaving the van for three days, we discussed our shared experience of teaching and mentoring novice researchers committed to working qualitatively but not knowing where to start. We thank our husbands, Tom Richards and Bob Morse, for contributing to these conversations, and for their support during the writing process. This text has also developed from our separate attempts to teach qualitative methods to students and professionals and to assist with their research. We thank our students, colleagues, and friends for their questions and challenges, for sharing their confusions and insights, and for providing opportunities to explain abstract and complicated concepts and techniques. At Sage, thanks to our editor, Lisa Cuevas Shaw; to her assistant, Karen Greene, for her enthusiastic help in making this edition happen; and to the copy editor, Cheryl Rivard, for her fine attention to its final detail.

Lyn Richards & Janice M. Morse

1

Why *Readme First?*

W hy *Readme First?* Why should a researcher, new to qualitative inquiry, begin by reading a book on the range of ways of doing qualitative analysis? Why not just start by collecting the data and worry later about what to do with them?

The answer is simple: In qualitative research, collecting data is not a process separate from analyzing data. The strength of qualitative inquiry is in the integration of the research question, the data, and data analysis. There are many ways of gathering and managing data, but because qualitative research is always about discovery, there is no rigid sequence of data collection and analysis. If you collect data and then select a method for analyzing them, you may find that the method you have chosen needs different data. To start with a *method* and impose it on a research question can be equally unhelpful. Good qualitative research is consistent; the question goes with the method, which fits appropriate data collection, appropriate data handling, and appropriate analysis techniques.

The challenge for the novice researcher is to find the way to an appropriate method. If the researcher new to qualitative inquiry evaluates the possible paths well and makes good choices, he or she can achieve a congruence of research question, research data, and processes of analysis that will strengthen and drive the project. However, this may seem to be an impossible challenge. The process of qualitative inquiry all too often appears as a mystery to the new researcher, and the choice of an appropriate method of analysis is obscured. The embattled researcher too often resorts to collecting large amounts of very challenging data in the hope that what to do with them will later become apparent. Some researchers end projects that way, still wondering why they were doing this or that or what was to be done with all the data they collected.

Readme First is an invitation to those who have a reason for handling qualitative data. We see qualitative research as a wide range of ways to explore and understand data that would be wasted and their meaning lost if they were preemptively reduced to numbers. All qualitative methods seek to discover understanding or to achieve explanation *from* the data instead of from (or in addition to) prior knowledge or theory. Thus the goals always include learning from, and doing justice to, complex data, and in order to achieve such understanding, the researcher needs ways of exploring complexity.

Qualitative data come from many sources (e.g., documents, interviews, field notes, and observations) and in many forms (e.g., text, photographs, audio- and videotapes, films). Researchers may analyze these data using very many, very different methods. But each method has integrity, and all methods have the common goal of making sense of complexity, making new understandings and theories about the data, and constructing and testing answers to the research question. This book is an invitation to new qualitative researchers to see many methods—to see them as wholes and as understandable unities. This makes the choice of method necessary but also makes the process of choosing enabling rather than alarming.

> In this book we use the term *method* to mean a collection of research strategies and techniques based on theoretical assumptions that combine to form a particular approach to data and mode of analysis.

This book provides the beginning researcher with an overview of techniques for making data and an explanation of the ways different tools fit different purposes and provide different research experiences and outcomes. Our goal is not to present a supermarket of techniques from which the researcher can pick and choose arbitrarily; rather, we aim to draw a map that shows clearly how some methodological choices lead more directly than others to particular goals. We see all qualitative methods as integrated and good qualitative research as purposive. Unless the researcher has an idea of the research goal, sees from the beginning the entire research process, knows the contents of the appropriate analytic toolbox, and recognizes from the start of the project what may be possible at the finish, it is not advisable for him or her to begin.

This book is not intended to be a sufficient and complete sourcebook, but rather a guide to doing a project. Indeed, it is intended be read before a researcher begins a project. The book is about how ways of collecting and making data are connected to ways of handling data skillfully, and how qualitative methods allow researchers to understand, explain, discover, and explore. It is our intention to inform readers of the research possibilities, direct them to the appropriate literature, and help them on their way to trying out techniques and exploring the processes of analysis. By informing themselves about the possibilities for analysis and the range of methods available, new researchers can critically select the methods appropriate to their purposes.

We wrote this book because as researchers, teachers, mentors, and advisers, we have suffered from a vast gap in the qualitative research literature. Most texts describe a single method, often not explaining how purpose, data, and analytic technique fit together. A few display the range of qualitative methods, but a novice researcher is seldom helped by such displays if they include no explanation of how and why choices can be made. The confusion is worsened if the researcher is led to believe that one method is required for reasons of fashion, ideology, alleged superiority, or pragmatic necessity (i.e., only this one method can be supervised or approved in the research site). A researcher may be caught between instructions for a particular method and research reports that offer no sense of how those who did the research got there. In this volume, we offer to bridge at least some of these gaps.

We also wrote this book because the present literature offers no picture of how to envision, at the beginning, the completion of a project. Researchers approaching qualitative inquiry need to be able to see the end before they start. In the chapters in Part III, we advise the reader on the goals to be aimed for, on rigor and reliability, and on the processes of finishing and writing it up. In the final two chapters, we deal with getting the reader started on his or her own project and smoothing the challenges of the start-up.

Readme First is neither a substitute for experience nor an instruction manual for any particular method. Researchers who want to use the techniques we describe here on their own data are directed to methodological literature that offers fuller instruction in particular methods. Nor do we intend this book to be a substitute for the new researcher's learning how to think qualitatively alongside an experienced mentor. We are both sure that qualitative research, like any other craft, is best learned this way. But many researchers do not have the opportunity to work with mentors, and sometimes the learning experience can be confining even while it is instructive. In this book, we present some practical ways new researchers can try out various techniques so that they may develop their skills. Exploiting these practical examples will give researchers insights into why they should use certain procedures and build their confidence to try them. Our goals are to demystify analysis, to promote informed choice, and to assist researchers in test-driving techniques while avoiding generalizing across methods or smudging the differences.

⸙ GOALS

In the development of this volume, we identified five related goals:

1. To emphasize the integrity of qualitative methods

2. To present methodological diversity as requiring informed choice

3. To demystify qualitative methods

4. To introduce qualitative research as a craft and to provide researchers with information on ways in which they can gain experience before launching their projects

5. To present qualitative methods as challenging and demanding

⸙ METHODS AND THEIR INTEGRITY

A strong message in this book is that although there is no one (or one best) approach to handling and analyzing qualitative data, good research is purposive and good methods are congruent with a fit among question, method, data, and analytic strategy. There are common strategies and techniques across all methods. It is these commonalities that make it sensible to talk about "qualitative methods." But techniques and strategies make methodological sense only in the context of particular methods, and it is the method that molds how the strategies and techniques are used. Therefore, although informed and debated innovation strengthens and changes methods, researchers do not gain by picking and choosing among techniques and incorporating them out of their methodological context.

Qualitative research helps us make sense of the world in a particular way. Making sense involves organizing the undisciplined confusion of events and the experiences of those who participate in those events as they occur in natural settings. Qualitative methods provide us with a certain type of knowledge and with the tools to resolve confusions. Behind the selection of method is often, but not always, an explicit or implicit theoretical framework that carries assumptions about social "reality" and how it can be understood. Various qualitative methods offer different prisms through which to view the world, different perspectives on reality, and different ways in which to organize chaos. Further, they use different aspects of reality as data, and it is the combination of these different data, different perspectives, and different modes of handling the data that gives us different interpretations of reality.

Because the method the researcher uses influences the form the results will take, it is imperative that the researcher be familiar with different kinds of qualitative methods, their assumptions, and the ways they are conducted before beginning a qualitative project. Such preparation will ensure that the researcher's goals are achieved, that the assumptions of the research have not been violated, and that the research is solid.

To argue for *methodological integrity* is not to argue for rigidity in methods. Methods rarely stay unchanged, and it is essential that they evolve over time. Researchers develop new techniques when confronted by challenges in their data, and if these techniques are consistent with the methods, they are drawn into other researchers' strategies. We both find excitement in methodological change and debate and have both been actively involved in it. However, we argue that innovations must be evaluated and critiqued within a method and developed with caution by seasoned researchers. Researchers who approach analysis by mixing and matching techniques derived from different methods without understanding them in their context commonly end up with a bag of techniques unlinked by strategies and uninformed by method, techniques that have nothing in common except that they are in the same project bag. Specifically, we want to warn researchers against using all the tools that particular computer programs provide without asking whether these techniques fit the research question, the research method, and the data.

〉〉〉 METHODOLOGICAL DIVERSITY AND INFORMED CHOICE

Our second goal is to display the diversity of qualitative methods and, in so doing, help the new researcher in choosing a method. As we noted above, the literature is dominated by texts that teach one particular way of doing qualitative research. Those texts are essential in that they provide the detail researchers need to work with particular methods, but in our experience, newcomers need an overview of the range of methods to help them envision the possibilities and outcomes of using alternative methods. Just as automobile manuals tell you little about the processes of driving, the menus in a software package do not tell you how to *analyze data* or how to use the software with different qualitative methods.

We begin with the assumption that no one method is intrinsically superior to others; each method serves a different purpose. For any given project or purpose, there may well be no method that is obviously best suited. However, the researcher needs to identify which method is *most* appropriate and then go to the relevant texts. Hence the title of this book. This volume is intended not as a substitute for the texts on particular methods, but rather as tool to help researchers access those texts. Like the README files that come with computer applications, it is intended to be read before the researcher commences the research process. We hope that researchers will be led from this book to particular methods and that what they learn here will help them to make informed choices concerning what they do during the research process.

We start with a sketch map of (some) qualitative methods. This particular methodological map may puzzle those familiar with the qualitative literature because it deliberately ignores disciplinary boundaries. We strongly believe that the development of qualitative methods has been retarded by narrow debates and the inability of many researchers to learn from, or even read about, the methods used in other disciplines. For instance, although ethnography was developed within anthropology (and often best answers questions asked by anthropologists), researchers from other disciplines (e.g., education) often ask ethnographic types of questions and are thus best served by ethnographic method. But research methods have been subject to waves of fashion, so that, for instance, in health sciences, the relevance of ethnography is often ignored in favor of other methods that may be less suited to particular projects, such as grounded theory or phenomenology. Disciplines do not "own" methods, and researchers are deprived of resources if they are prevented from looking beyond the current trends in their own disciplines.

Our methodological map is designed only for orientation; it is not complete, and it gives relatively little detail. We do not attempt to map all forms of qualitative inquiry; rather, we want to distinguish major methods in order to show and encourage methodological diversity, integrity, versatility, and respect for the many ways of making sense of data and making theory from data.

∭ NO MYJTERIEJ!

Our third goal is to demystify qualitative methods. Each method provides a cluster of approaches or techniques to use with data—techniques requiring plenty of skill but no magic. New researchers who are awed by the great mysteries of analysis are inhibited from trying their hands at making sense of data, even when they urgently wish to do so.

Demystifying is always dangerous, risky, and trivializing. Good qualitative research certainly summons—and deserves—amazement, awe, and excitement for the complex processes involved in constructing new understandings and arriving at explanations that fit. We do not intend our discussion in this book to remove that excitement. But we see qualitative research as a craft, not a mystery, and as cognitive work, not miraculous and instantaneous insight. The processes of good qualitative analysis are exciting—not because they are mysterious, disguised by the wave of the magician's wand, but because, like the work of the sculptor, they are the result of skilled use of simple tools, practiced techniques, focus and insight, concentrated work, and a lot of hard thinking.

This book, then, is about agency. Researchers make data and work with data as they attempt to derive from them accounts and theories that satisfy. We offer no "black box" from which theory "emerges." To do justice to qualitative analysis, researchers have to be able to see how messy data can be transformed into elegant understanding and that this is something that can be attained by normal folk. In this, they will be helped by practical accounts of how it has been done and hindered by passive-voice accounts of how themes are "discovered" and assertions that a theory "emerged." We believe good qualitative research requires that not only researchers be actively involved in data making and interpreting but that they account for and describe their progressive understanding of their data and the processes of completion. This is an active and intentional process, one that researchers control, develop, shape, and eventually polish. It is therefore enormously exciting and rewarding.

∭ LEARNING BY DOING IT: QUALITATIVE REJEARCH AJ A CRAFT

Like any craft, qualitative research is best learned by doing it and talking about the experience. We have learned that teaching qualitative methods in abstraction, without involvement in data, works for very few students. Yet most introductory texts offer rules rather than experiences. Our fourth goal is to offer learning by doing. In this book, we offer few rules; instead, we offer many explanations of techniques and the way they fit methods as well as suggestions for test-driving the techniques discussed. To take into account the wide variety of researchers' interests, the CD-ROM that accompanies the book includes several samples of qualitative data.

Of course, we cannot attempt to teach all of the aspects of the major craft of qualitative analysis, with its long history and rapidly changing techniques, in this small volume. Our goal is to give a sense of what competent qualitative craftsmanship can do to data. Therefore, this book is not a "dummy's guide"; we do not provide abbreviated instructions that result in trivial projects. We do not spoon-feed readers, and we do not give instructions regarding sequential steps they should take. Rather, we offer readers ways of exploring the aims and effects of the central qualitative techniques and of getting a sense of what these techniques do to data in the context of particular methods.

To see qualitative research as a craft is to resist trends toward atheoretical qualitative inquiry—inquiry that is not grounded in a theoretical context or that merely reports selected quotations. Qualitative research is an intellectual activity firmly based on the cumulative intellectual activities of those who have come before and their respective disciplines. Therefore, in this book, we aim not only to assist researchers in trying out techniques but to help them see those techniques as making sense in the context of a given method with a theoretical framework, a history, and a literature.

We tackle this goal with attention to the software tools currently available for handling qualitative data. These are changing rapidly, and we share a concern that technological advances should not further obscure or replace the craft of analysis. Whether researchers handle their data using index cards or sophisticated software, the essential first step is to learn to think qualitatively. When handling the data is done with software, the researcher must understand that software does not provide a method.

Selection of *some* tools for doing analysis requires an understanding of how analysis might be done with *other* tools. It is now common for researchers to use specialized software tools for at least some qualitative research processes, but the qualitative methods literature has handled the discussion of computer techniques poorly, if at all, and researchers need to know what computers can and cannot do. Computer programs may come to dominate the ways researchers handle data and probably have contributed to the explosion of qualitative research. Yet novice researchers often see such programs as offering a method. For that reason, this book will look at what qualitative researchers can and cannot do with computers.

An overview of what software does is provided in Chapter 4, to assist you in choosing the appropriate software tools for your project. Four tables summarize what all programs do, then what variety there is and when this will matter. They are designed to help the researcher to see software choice, like methods choice, in terms of the requirement for methodological congruence.

Each of the chapters about techniques of handling data (4, 5, 6, and 7) and writing up your study (10) concludes with a summary of what you can expect from your software and advice and warnings to help you use it well. Qualitative software tools are developing rapidly, and the software in turn changes methods, since it allows researchers to handle data and ideas in ways not feasible without computers. So these chapter sections do not describe the range of current software. Any printed account of particular functions of available software would be immediately out of date. To learn about the range of qualitative software available to you, and the functions and tools that different software packages offer, you need to go to websites. This is easily done via the University of Surrey's CAQDAS Networking Project, whose website provides up-to-date summaries of current software and links to the websites of all qualitative software developers: http://www.soc.surrey.ac.uk/caqdas/.

The sections on software in this book offer something different. Rather than comparing current software functionality, they explore the ways qualitative work is supported and changed by software tools. They are written by Lyn Richards, who was part of the development team for two software packages, NUD*IST and NVivo. But the chapter sections do not refer to or teach any particular software. For those who wish to try out these tasks with current software, there are tutorials in NVivo accompanying Richards's (2005) companion book, *Handling Qualitative Data: A Practical Guide,* online at www.sagepub.co.uk/Richards.

〰 QUALITATIVE REJEARCH AJ A CHALLENGE

Our fifth goal concerns the public relations of various qualitative methods. We confront the widespread assumption that qualitative research is simple and that to "do qualitative" is easier than conducting quantitative research, because you don't need statistics and computers; we also deal with the attendant assumption that these are "soft methods for soft data." We present qualitative methods as challenging and demanding, made so because they can (and must) be rigorous and can (and should) lead to claims for conclusions that are defensible and useful.

The challenge is not in doing it "right," but in achieving coherent, robust results that enhance understanding. We present our readers with principles rather than hard-and-fast rules to be followed. We conclude the book by addressing the issues of rigor and the ways in which it is achieved, assessed, and demonstrated.

There is also a challenge in reconciling the sometimes opposing requirements of different methods. We emphasize, rather than obscure, what we consider to be the essential paradoxes inherent in qualitative research. Central among these paradoxes are the opposing requirements of simultaneous pursuit of complexity and the production of clarity.

We explore and discuss the built-in contradictions that texts often submerge, dodge, or totally ignore. It is our experience that novice researchers find sometimes insurmountable barriers in unexplained paradoxes. Too often, they are left puzzled and paralyzed, feeling responsible for their inability to progress toward analysis. If understood as an integral part of analysis, however, these are challenges, not barriers. Meeting these challenges is a normal and necessary part of coping with complex data. Confronted, they offer hurdles that can and must be cleared, and all qualitative researchers know the pleasure of clearing such methodological obstacles. Once a researcher has acquired the proper tools, these obstacles become exciting challenges rather than reasons for giving up.

⊗ USING README FIRST

Warning: This book is designed to be read like a novel—it has a story. If you skip a section, later parts may not make sense; our best advice is that you skim read before you jump in fully.

Terminology

We use specific terms in specific ways. When we use the term *method*, we refer to a more or less consistent and coherent way of thinking about and making data, way of interpreting and analyzing data, and way of judging the resulting theoretical outcome. Methodological principles link the strategies together. These methods are clearly labeled and have their own literatures. Examples of such methods are ethnography, grounded theory, and phenomenology. However, methods vary in their completeness, and a great amount of qualitative research is not done within traditional methods. We share a concern that researchers feel coerced to stick a traditional label on less complete methods. In later chapters, we describe the complete methods that are less coherent and not fully developed and the challenges involved in using them.

A research *strategy* is a way of approaching data with a combination of techniques that are ideally consistent with the method the researcher has chosen to use. Strategies, therefore, are based on, and are consistent with, the assumptions and procedures that are linked in each particular method. We will argue that strategies made up of techniques that have been haphazardly and arbitrarily selected from different methods are problematic.

We also use the term *technique* to refer to a way of doing something. In our context, research techniques are ways of attempting or completing research tasks. If you see someone using a particular technique (for example, coding data), that technique might not tell you which method the researcher is using—everyone codes data. But if you look more closely at the ways in which the researcher is applying that technique and at where it takes the researcher, you will be able to determine the method the researcher is using. Coding does different things to data when it is done by researchers using different methods.

We aim to map commonalities while explaining diversity and to present methodological techniques in ways that will help researchers to arrive at coherent strategies within understood methods. Our overall goal is to help readers develop a sense of methodological purpose and appropriateness, and at the same time, we provide an evaluation and critique of qualitative research. We hope that this book will help those readers who go on to do their own research to know what they are trying to do and why they are doing it one way rather than another way; we want to help our readers to achieve the most satisfying answers to their research questions, the best sense of discovery and arrival, and the best new understanding with the most efficiency and expediency.

⧓ THE ſHΛPE OF THE BOOK

We begin by establishing the integrity of methods and then approach, in turn, the different dimensions of qualitative research that researchers have to understand in order to be able to start their own research projects. In each section, we aim to give an idea of how it would be to work in a specific way.

At the end of each chapter for which software skills are relevant, we discuss how it will feel to work with software, and we advise on the use of computer tools. We provide web links to give up-to-date information about current software and to tutorials for learning a software package.

Each chapter concludes with a list of resources to direct readers to the literature on each of the methods discussed, to wider literature, and to completed examples of relevant research. This literature deals with the processes of thinking qualitatively, preparing for a project, relating to data, and creating and exploring ideas from the data and theories about the data.

The chapters in Part I, "Thinking Research," address our first two goals: to establish the integrity of qualitative methods and to present methodological diversity as a choice, not a confusing maze. They provide a view from above, to be used as one would use an aerial photograph to scan a particular terrain and understand possible routes to a given destination.

In Chapter 2, which deals with the integrity of qualitative research, we set out what we see as core principles—the *purposiveness* of method and the methodological *congruence* of qualitative research. We show how different methods fit different sorts of qualitative data and how they have different implications for analysis. This very general overview informs the discussion in later chapters about the range of ways of meeting and handling data and the range of analysis processes and outcomes.

We compare three methods in Chapter 3 as we present the case for methodological congruence. These are the methods most widely used: phenomenology, ethnography, and grounded theory. Although there are many variants within each method, each is identified by characteristic ways of addressing questions through data. Each method is appropriate to particular types of questions, each directs researchers to make particular research designs and data, and each leads researchers to use particular techniques for handling data and discovering and analyzing meanings.

Chapter 4 is about research design, and it has a simple message: A researcher absolutely needs a research design. We discuss why design is often demoted or ignored in qualitative research and urge that researchers take the opposite approach. Like the methods they express, research designs should not be seen as fixed or holy. However, careful consideration and planning set a project on the path to its intended goals and maximize the likelihood of getting there. We explore what researchers can and cannot plan, and we emphasize the design of the *scope* of the project and the *appropriateness* of the data.

The chapters in Part II, "Inside Analysis," are concerned with what it is like to be doing qualitative research. Chapter 5 is about data: the range of ways of making data, the role of the data at the beginning of the project, the sources and styles of qualitative data, data required for different methods, and when data will be useful and when not. We emphasize the agency of the researcher in making data collaboratively with "subjects," and the ways data are crafted to meet the research goal from the beginning of the project.

All methods share the goal of deriving new understandings and making theory out of data. But novice researchers are often unable to get a sense of the research experience behind these goals. What is a category? How would I know one if I found one? What should coding *do* for you? What is it *like* to create theory? In the remaining chapters in Part II, we discuss and demonstrate the tools for handling and coding data and for theorizing. Starting with abstraction, we move to the common processes of using and developing categories and linking them to data through coding. In Chapter 6, we examine the central and varied processes of coding and the different ways in which researchers can use coding to move between data and ideas. Chapter 7 deals with the goal of abstracting and "theme-ing," or "thinking up" from the data, which is common to all methods. In Chapter 8, we address the goals of theory and theorizing, discussing different ways to reach those goals. We tackle the mysterious processes behind the claim that theory "emerges" from the data and show the ways researchers create and evaluate the fit of theory and data. Our emphasis is on the varieties of theory, on the processes of recognizing theory and working it, and on the layering of theory and building it up.

The chapters in Part III, "Getting It Right," are concerned with the process of completing qualitative analysis so that it works for the researcher's purposes. In Chapter 9, we discuss getting analysis right as well as the ways researchers can know if it is wrong. Chapter 10 deals with reporting results and writing them up, ensuring that a qualitative project will be credible and persuasive, and ways in which researchers can aim for these goals.

Chapters 11 and 12 in Part IV, "Beginning Your Project," end the book by describing the groundwork researchers need to do to begin their own projects and get into the field. We recommend that while researchers wait for the permissions they need to begin their projects, they "get skilled" by selecting and learning to use appropriate software for their analysis. We finish with encouraging words to get the new researcher started.

Appendix 1 is a guide to the reader wishing to use software tools along-side each of the central chapters. It specifies the relevant tutorials on the associated website. Appendix 2 discusses how to apply for funding.

〟 WHAT TO EXPECT

This book is intended to be read at the inception of a project and reread as needed until writing is completed. We recommend that you consult it when you wonder why you are doing this or that or where your current path is taking you. Qualitative researchers differ greatly from quantitative researchers in the way they approach research. Usually, qualitative researchers start with *areas of interest* or general, rather than specific, research questions. They may not know very much about the topic at the start, and even if they do, they seek to learn more through the data. Similarly, you must be flexible. You need to start with a broad understanding about the general area, be receptive to new ideas and willing to relinquish old—but unsupported—favorite ideas, and obtain a notion of the boundaries from the phenomenon studied. In all qualitative methods, one goal is to create categories and linkages systematically from the data, confirm these linkages, and create theory. You will find that it is easier to achieve this objective if you understand the entire research process and have an overview of the entire project, knowing what steps come next.

If you are approaching qualitative research with no idea what it will be like, here is a sketch. It is not, of course, a picture of an ideal project (or any real project), but an impression of the ways things tend to develop. It gives a simple overview of the research process and the ways in which you might interface with *Readme First*. If this sketch were to represent reality, it would be a mess of loops and double-headed arrows—qualitative research is more often cyclical than linear. But although you cannot expect a tidy procession of stages, qualitative research usually has some predictable progress; during most projects, there are series of periods during which a few things happen simultaneously. We revisit this picture in the final chapter.

Even before selecting a research topic, you must understand the nature of qualitative methods. You must know what qualitative methods *can do* and *cannot do*, where and for what kinds of problems and questions they should be used, and what kind of information is obtained through the use of various qualitative methods. We start with this point in the next chapter.

The process of learning to think qualitatively—to think like a qualitative researcher—can be challenging. If you do not have training, our best advice is that you read basic introductory texts, take an introductory course, talk to researchers about their experience and read their studies, and find and read *critically* a wide range of published works by researchers who have employed different qualitative methods. Such a broad overview will give you a feel for the field. Ask yourself: What kinds of questions are best answered using qualitative methods? What kinds of qualitative methods are best used with certain questions? What is the relationship between the data and the emerging results? What does "good" research look like? In particular, you should explore how research results vary in their level of theoretical development, from simply reporting and organizing quotations to creating sophisticated and elegant theories. Ask yourself why some research seems satisfying and some less so. You should be asking all these questions simultaneously.

Becoming Focused

Read *Readme First.*

Learn to think qualitatively.

Read other texts, take a course, and talk to researchers.

Reflect on, refine, and define a topic area.

Start to shape a research question.

⟫ DOING QUALITATIVE REJEARCH

A qualitative researcher usually locates a *topic* first—not a specific question and very rarely a research location or sample. This is not a methodological or moral imperative, but if you start, for example, with a group you wish to study, you would be well advised to broaden your vision to a wider context. You should resist the temptation to move directly to research design or, worse, to make a list of the questions you are going to ask your study participants.

If you approach the topic from a broad perspective, it will lead you to the literature. You should examine and analyze other studies critically, both within the context of the proposed research and within the context of the researchers' disciplines. But, most important, you should examine the literature *qualitatively.* It is not enough to summarize or synthesize others' results. Rather, you need to examine the theoretical perspective and method of each study, looking for overt and covert assumptions, beliefs, and values that contributed to the researcher's perspective, questions, selection of hypotheses, and interpretations of results. For a while, you should combine these tasks.

Becoming Competent Methodologically

Read extensively around the topic.

Read extensively on the possible methods.

Develop and learn the ways you will handle data.

Narrow down your methodological options.

Choose your software and learn it.

Such a critical appraisal of the literature is a student's first step in qualitative inquiry—and in qualitative analysis. This may also be your first step in handling qualitative data. You should treat the literature review as a data-managing exercise. As you work through this book, consider how each method we discuss can be applied to making sense of your reading (which, just like interview data, builds up in unstructured text records).

Now is the time to start *managing data* skillfully. If you are planning to use a computer program to handle your data, learn it *now* and use it to organize your notes and any discussions arising from the literature. Things move fast once you have located your study methodologically, and competence with your software will help you maintain the pace and maximize the exploration of data as they accumulate.

Your new understanding of the literature, and the acquired understanding of qualitative methods in general, will direct you toward the research question, the appropriate qualitative method, and, thus, the start of a research design. Resist the temptation to narrow the research question too far; you will refine and delimit it during the process of data collection. Resist any pressure to select your method until you are sure you know where your study fits.

Shaping the Study

Locate the study methodologically.

Locate the study in the research field.

Work on and rework a research design.

Start making some appropriate data.

Start data analysis *now.*

Manage data and ideas.

Every researcher experiences this stage as a flurry of activity and impending chaos. Reading on method is imperative if you are to be sure-footed in your entry into the research field. The importance of learning to think theoretically will be evident as soon as you begin data collection. Any observation or piece of text can be seen in two ways: It can be taken at face value, or it can be viewed as theoretically rich, linked to other pieces of data, linked to existing theory, and linked to your ideas. Our best advice is that you take this stage of interlocking tasks carefully and slowly. Never allow the excitement and the demands of the impending project to distract you from *designing* your research.

It is important at this point that you develop a systematic and simple means of documenting, linking, sorting, and storing these ideas. The system must be fluid, so that the developing codes and categories remain malleable as the ideas change and evolve as your comprehension increases. If you are using a computer program, talk to other users and partake in online discussions to gain a sense of what tools the program offers you and which ones you can use.

Note that you have now commenced analysis work: active, hard, deliberate cognitive work. You are not mindlessly gathering data as if picking apples; your analysis should be ongoing and never delayed until all data are in. If you are working qualitatively, it is the data-driven analysis that will tell you when the data are adequate.

Conceptualizing and Theorizing

Actively seek theory.

Constantly check with data.

Explore complexity and context.

Simplify and integrate.

Sift, sort, and play with data.

Processes of making data and making the analysis continue. Your early ideas and data sortings look simplistic, but the "right" solutions often appear beyond your grasp. Although this is an intriguing and exciting stage, it can also be the most frustrating and most difficult one. Return to this volume for overviews of the data-handling and theory-generating processes. Ensure that you keep analyzing as you make data and that you allow the data to direct you to ideas that surprise you and that you hadn't previously thought to explore.

Explore your data from different perspectives. Play with your data. Pursue hunches and think outside of the tidy explanations. Write, write, write, and rewrite. Create models and discuss them. Confirm ideas in your data or collect additional data. Discuss your theories with anyone who will listen. Compare the emerging theory with the theories in the literature. And, most important, think! Consider the research as a puzzle to be solved, a solution as always possible, and the process as active mind work. Theory does not emerge overnight; data never "speak for themselves."

Molding and Writing

Arrive at a best account of the data or theory to make sense of the data.

Tidy up and polish.

Write, present, and publish.

It may happen suddenly that all your research will come together and integrate in a flash of discovery, or it may happen slowly over a period of time. But eventually your research will make sense. The growing web of ideas and theory will be strong enough to support a story, an account, or an explanation that makes sense of the data. Your familiarity with the literature will have given you a sense of the final product, but perhaps not of something achievable by you. Like all extraordinary experiences, it will be different from what you expected, and you will be astonished when it happens. You can *tell* your study. You have arrived at a solution—a beautiful, elegant solution—that is supported with data, connects with the literature, and makes sense in the research context. Your study, *if* tidied up and polished, will make an important contribution to the literature.

Keep the momentum going until your study is published and accessible to all. And when that is done, with great pomp and ceremony, give *Readme First* to a friend.

※ RESOURCES

Read widely among the available basic texts to get a feel for how to approach qualitative analysis.

Major Resources

Creswell, J. W. (1998). *Qualitative inquiry and research design: Choosing among five traditions.* Thousand Oaks, CA: Sage.

The five traditions covered in this text are biography, phenomenology, grounded theory, ethnography, and case study.

Denzin, N. K., & Lincoln, Y. S. (Eds.). (1998). *The landscape of qualitative research: Theories and issues.* Thousand Oaks, CA: Sage.

This book is a reprint of the methods section of the first edition of the *Handbook of Qualitative Research*. It contains summaries of the major qualitative methods written by leading experts or the developers themselves.

Denzin, N. K., & Lincoln, Y. S. (Eds.). (2000). *Handbook of qualitative research* (2nd ed.). Thousand Oaks, CA: Sage.

This and the third edition contain updated chapters from the first edition and many new chapters. A quick tour of many approaches.

Denzin, N. K., & Lincoln, Y. S. (Eds.). (2005). *The Sage handbook of qualitative research* (3rd ed.). Thousand Oaks, CA: Sage.

The 3rd edition of the series contains new chapters. An important reference book.

LeCompte, M. D., Millroy, W. L., & Preissle, J. (Eds.). (1992). *The handbook of qualitative research in education*. San Diego, CA: Academic Press.

This volume provides a comprehensive overview of major qualitative methods, particularly for those in education.

Mason, J. (2002). *Qualitative researching*. London: Sage.

This text gives an overview of qualitative methods and clear discussion of many of the current issues students confront.

Richards, L. (2005). *Handling qualitative data: A practical guide*. London: Sage.

This is a companion work to *Readme First*, the present book, advising on what to do when you have data. It gives detailed advice for doing the tasks and techniques described in the next chapters. (See also the software tutorials at http://www.sagepub.co.uk/richards/ and resources at http://www.lynrichards.org/.)

Schwandt, T. A. (1997). *Qualitative inquiry: A dictionary of terms*. Thousand Oaks, CA: Sage.

This detailed and handy reference book provides definitions of the many new terms novice researchers encounter as they delve into qualitative inquiry.

Additional Resources

Berg, B. L. (1998). *Qualitative research methods for the social sciences* (3rd ed.). Boston: Allyn & Bacon.

Bernard, H. R. (1994). *Research methods in anthropology: Qualitative and quantitative approaches*. Walnut Creek, CA: AltaMira.

Bernard, H. R. (2000). *Social research methods: Qualitative and quantitative approaches*. Thousand Oaks, CA: Sage.

Crabtree, B. F., & Miller, W. L. (Eds.). (1999). *Doing qualitative research* (2nd ed.). Thousand Oaks, CA: Sage.

Creswell, J. W. (2003). *Research design: Qualitative, quantitative, and mixed method approaches* (2nd ed.). Thousand Oaks, CA: Sage.

Erlandson, D. A., Harris, E. L., Skipper, B. L., & Allen, S. D. (1993). *Doing naturalistic inquiry: A guide to methods*. Newbury Park, CA: Sage.

Ezzy, D., Liamputtong, P., & Hollis, D. B. (2005). *Qualitative research methods*. Oxford, UK: Oxford University Press.

Flick, U. (1998). *An introduction to qualitative research*. London: Sage.

Grbich, C. (1999). *Qualitative research in health: An introduction*. Sydney, Australia: Allen and Unwin.

Marshall, C., & Rossman, G. B. (1999). *Designing qualitative research* (3rd ed.). Thousand Oaks, CA: Sage.

Mason, J. (1996). *Qualitative researching*. London: Sage.

Miller, G., & Dingwall, R. (Eds.). (1997). *Context and method in qualitative research*. London: Sage.

Morse, J. M., & Field, P. A. (1995). *Qualitative research methods for health professionals* (2nd ed.). Thousand Oaks, CA: Sage.

Prasad, P. (2005). *Crafting qualitative research: Working in the post positivist tradition.* Armonk, NY: M. E. Sharpe.

Ritchie, J., & Lewis, J. (Eds.) (2004). *Qualitative research practice: A guide for social science students and researchers.* Thousand Oaks, CA: Sage.

Rossman, G. B., & Rallis, S. (1998). *Learning in the field: An introduction to qualitative research.* Thousand Oaks, CA: Sage.

Rothe, J. P. (2000). *Undertaking qualitative research: Concepts and cases in injury, health, and social life.* Edmonton: University of Alberta Press.

Seale, C., Gobo, G., Gubrium, J., & Silverman, D. (2004). *Qualitative research practice.* London: Sage.

Part I

THINKING RESEARCH

2

The Integrity of Qualitative Research

When commencing a qualitative research project, it is essential that the researcher understand the variety of methods available and the relationships among research questions, methods, and desired results. In this chapter, we try to show how a researcher choosing a topic is led to a method, what is possible for the research to achieve, what the researcher can ask and hope to have answered, and how question, data, and analysis fit together. Once a researcher sees this fit, the choice of a method for any particular study is never arbitrary.

Not all qualitative methods integrate all aspects of the project in the same manner, and most contain considerable variety. In this overview, we ignore those variations to stress the two principles of qualitative methods that inform the rest of this book: methodological *purposiveness* and methodological *congruence.*

First, we establish how particular research *purposes* and questions lead the researcher to particular data sources and analysis strategies, sketching the links for three major methods. We then argue for the importance of *congruence*—the way in which what the researcher asks, where he or she asks it, and how he or she works toward an answer all fit together.

※ METHODOLOGICAL PURPOSIVENESS

There is almost always a best way to do any research project, a particular method that is best suited to each particular problem. The choice of best method always comes from the research purpose.

Of course, the choice is never entirely open. It is always constrained by something—the researcher's familiarity with methods, the researcher's resources, or sometimes the data themselves. At one extreme, researchers starting from the availability of particularly interesting data will quite normally have their methodological options predetermined. Although this can be restricting, such researchers may well be envied by other researchers for whom none of the elements in the equation are controlled: a general topic area, many possibilities for making data, and no methodological direction. Researchers in the latter group, in turn, may be tempted to claim constraint ("I have to do a grounded theory study because that's the only sort of qualitative research accepted in my school"). But that's where the danger lies—in a topic shoehorned into a particular method. Some seasoned researchers work the other way around, through commitment to one method, which means they ask only (even, it might appear, *can* ask only) certain sorts of questions. But they start with questions, and they must always be open to the possibility that a question requires a different method.

Especially when choice of method seems constrained, it is important that the researcher understands the process by which he or she selects a method, and that the researcher sees the selection as *deliberate* and as reflecting research purpose. The purpose may be to learn about a specific problem (e.g., "Why do residents not use the facilities?") or to understand a situation ("I wonder what the experience of . . . is"). Or the purpose may be no more specific than to learn more about a particular topic or to do justice to those interesting data that suddenly became available. In such a project, exploring the literature and spending time in the setting will help the researcher to focus on a clearer problem and frame a sharper research question, and the data will direct further inquiry. In any project, a decision about method does not just happen by default. A purpose, however unspecific, guides the researcher to a more focused research question and hence to a choice of method.

The researcher actively creates the link between purpose and method through a process of reflecting on purpose, focusing on a researchable question, and considering how to address it. That link is never, of course, a simple one-way causal connection. Most projects commence with an opening phase in which the researcher conducts an *armchair walk-through* and considers several routes and several methodological vehicles. The appropriate approach may not be a qualitative method. Sometimes the research purpose opens out to several research questions, each requiring a different qualitative method, or the interplay of qualitative and quantitative methods. But, however it is arrived at, the link of purpose to method is what gets a project going.

Why Are You Working Qualitatively?

Why did you select a qualitative method? Often, the researcher has a very practical goal for beginning the project—it may be an unanticipated problem area in the classroom or a particularly puzzling patient situation that the experts seem unable to understand. It may be an area in which patterns of behavior are statistically clear (changes in the birthrate, for instance), but researchers can only guess at reasons for these patterns without an understanding of people's own accounts of their behavior. It may be a policy area (such as urban planning) where the best-laid plans are thwarted by apparently irrational choices (incredibly, the slum dwellers didn't want to be relocated!). In each of these cases, the researcher chose to work qualitatively, with complex unstructured data from which new understandings might be derived. Below, we summarize the major reasons for working qualitatively—the research question requires it, and the data demand it.

The Research Question Requires It

For many of us, the first really good moment in a project occurs when we see how the research purpose can be pursued by one but not another means. In retrospect, this may be blindingly obvious. For instance, you need to understand what children *mean* to parents in this society before you can predict fertility rates, so what you must do is listen to parents' stories of parenthood rather than ask predetermined questions about birth control. The only way of making sense of classroom problems is to get an understanding of the latent processes of power—observe, listen to what is said in the classroom and the staff room, and examine the words and their meanings rather than simply distribute a questionnaire. What if the apparently irrational behavior of slum dwellers makes sense to them? The only way to find out is to hang around and observe their daily life, rather than assume that the condition of their housing is their top priority. Each of these purposes points toward one of the methods we sketch in Chapter 3.

Researchers who are brought (sometimes kicking and screaming) to a qualitative method driven by the topic often combine qualitative with quantitative methods. They may recognize their need to understand and to develop meaning prior to or subsequent to, rather than instead of, a quantitative study. Perhaps they require a larger-scale inquiry or systematic testing of hypotheses. In such situations, a qualitative component may precede a quantitative project and provide different types of findings for richer and more complete results. (We discuss such combinations of techniques when we address triangulated design in Chapter 4.)

The Data Demand It

It may be, however, that you have no such research purpose directing you to working qualitatively. What, then, has led you to such methods? A powerful push can come from data; some data can be obtained only through the use of a particular strategy. For example, it is not possible to interview some participants—very young children who cannot talk or elderly persons with Alzheimer's disease may not be able to provide coherent responses. In these cases, the nature of the participants requires that researchers use observational strategies, obtaining data in the form of field notes or videotape. If your topic forces you in such a direction, it will be the first of many times in the project when data seem to be driving the study. Recognizing such imperatives will always take you forward, because qualitative methods are properly responsive to discoveries in data.

Many quantitatively trained researchers first started working qualitatively because they recognized that the statistical analyses of particular survey responses did not seem to fit what those in the situations of interest said or what people wrote in their open-ended answers. In avoiding the temptation to dismiss their participants' open-ended responses or to use them merely to illustrate the reports, perceptive researchers sought ways to analyze them. Action researchers might be brought to qualitative methods by complex social or political situations in which it is essential to understand all sides of a controversy but the available documents and discussions defy neat categorization. For a study to be useful, the researcher must make sense of such a situation. Practitioners might observe and record the complexities of clinical situations that seem to be denied by tidy reports of patient compliance; in seeking an understanding of that complexity, they find they need ways of doing justice to the data.

Coming to a qualitative method because your data require it provides high motivation but often high stress, too. The survey must be reported, the action group informed, the patients helped; it seems that you must become an instant qualitative researcher. If this is your situation, we recommend that you go carefully through the nine points we list in Chapter 12 under the heading "How Do You Start?"

Should You Be Working Qualitatively?

The obvious first question is whether the research purpose is best answered by qualitative methods. We hope we have made it clear that we see nothing morally or methodologically superior about qualitative approaches to research. Other things being equal, a quantitative project will often be faster, easier for a researcher lacking qualitative training, and arguably more acceptable in many research contexts. Moreover, the research world is replete with questions that are properly and effectively answered quantitatively, and that will be badly answered, or not answered at all, if a qualitative method is imposed on them. Forcing such questions into qualitative methods has the same effect on projects and researchers as forcing the glass slipper onto their feet had for Cinderella's ugly stepsisters' marriage prospects—it won't work, it will hurt a lot, and the result is a loss of credibility.

It is not our goal in this book to examine the philosophical origins of qualitative methods or the approaches to evidence and "reality" behind different methodologies, but it is important to note that we see no chasm between qualitative and quantitative techniques. It is our experience that many qualitative projects involve counting at some stage, and many questions are best answered by quantification. But given that we aim here to give those embarking on qualitative research an understanding of what it will be like, we assume that you, the reader, are about to embark. Thus the obvious first question is whether you should do so.

Qualitative methods are the best or only way of addressing some research purposes and answering some sorts of questions, such as in the following cases:

1. If the purpose is to understand an area where little is known or where previously offered understanding appears inadequate (thin, biased, partial), you need research methods that will help you see the subject anew and will offer surprises. Put bluntly, if you don't know what you are likely to find, your project requires methods that will allow you to learn what the question is from the data.

2. If the purpose is to make sense of complex situations, multicontext data, and changing and shifting phenomena, you need ways of

simplifying and managing data without destroying complexity and context. Qualitative methods are highly appropriate for questions where preemptive reduction of the data will prevent discovery.

3. If the purpose is to learn from the participants in a setting or a process the way *they* experience it, the meanings they put on it, and how they interpret what they experience, you need methods that will allow you to discover and do justice to their perceptions and the complexity of their interpretations. Qualitative methods have in common the goal of generating new ways of seeing existing data.

4. If the purpose is to construct a theory or a theoretical framework that reflects reality rather than your own perspective or prior research results, you may need methods that assist the discovery of theory in data.

5. If the purpose is to understand phenomena deeply and in detail, you need methods for discovery of central themes and analysis of core concerns.

Each of these suggestions has a flip side. If you know what is being hypothesized and what you are likely to find, if you do not need to know the complexity of others' understandings, if you are testing prior theory rather than constructing new frameworks, or if you are simply describing a situation rather than deeply analyzing it, it is possible that you should not be working qualitatively. Perhaps the research question you are tackling with in-depth interviews would be more properly addressed with a survey. In such a case, our best advice is that you review your general purpose and ask yourself if it can be addressed better that way. Many purposes are perfectly served by survey data, and very many purposes require surveys. Important examples are research questions seeking to establish the associations among easily measured factors across a group or setting. If your goal is to establish that women in the paid workforce use neighborhood services less than do women who don't work outside the home, a survey will do it. But maybe what you really need to ask is how women in the paid workforce perceive neighborhood relations.

Or perhaps the research purpose can be addressed through the use of more straightforward techniques, such as quantitative content analysis. If you wish to know which words dominate discussions of medical treatments, rather than the meanings the participants give those words, a qualitative approach is likely to delay your answer. But maybe you want to find out more—for example, maybe you want to discover whether there are dominant discourses underlying those discussions.

On reflection, in either of the above cases there might be aspects of the research topic that would be best addressed through a combination of qualitative and quantitative data. As we will show in Chapter 4, such combinations fit easily with many qualitative methods.

Qualitative research is a proper response to some, but not all, research needs. We have both learned to be alert to risk in projects where the researcher is working qualitatively for the wrong reasons. These include reasons that are negative rather than positive ("I hate statistics" or "I can't use computers") and assumptions that qualitative research is more humanistic, moral/ethical, worthy, feminist, radical, or admirable. (The techniques we describe in the chapters that follow are also the most invasive, intrusive, and morally challenging; the only good reason a researcher should consider using them is that the research problem requires them.) Our point here is not just that you need a good reason for working qualitatively because of both practical and ethical considerations, but also that you need to have thought your way to this method if you are to start learning it. Good qualitative research requires purpose, skill, and concentration, and unless you recognize this and your purpose is clear and committed, the task will quickly become onerous.

How Should You Be Working Qualitatively?

What we have described as a fit between research question and method is never a simple cause-and-effect relationship. As you decide on the focus and scope of your study, the firming research question will indicate the best method for you to use, and your reading on methods will suggest ways in which you can focus the study. We start in Table 2.1 by comparing just three of the qualitative methods commonly described in textbooks. Usually (but not always), phenomenology best addresses a question about meaning: "What is the experience of . . .?" Ethnography offers researchers tools to answer questions such as "What is happening?" Researchers are directed to grounded theory by questions of interaction and process: "How does one become a . . .?" The link between question and data is obvious when one contrasts these three "classic" methods.

Table 2.1 The Fit of Question and Method

Type of Question	Method That Might Be Appropriate
Quesions about meaning (e.g., What is the meaning of . . . ?) and about the core or essence of phenomena or experiences	Phenomenology
Observational questions (e.g., What are the behavioral patterns of . . . ?) and descriptive questions about values, beliefs, and practices of a cultural group (What is going on here?)	Ethnography
Process questions about changing experience over time or its stages and phases (e.g., What is the process of becoming . . . ?) or understanding questions (e.g., What are the dimensions of this experience . . . ?)	Grounded theory

From Selecting a Method to Making Data

As the purpose points to the research question and the research question informs the choice of method, so the method fits the type of data to be collected. (See Table 2.2, which lists the types of data required by particular methods.) However, selecting a method and making data are not discrete events in the research process; rather, they are aspects linked by common ways of thinking. The distinction between a method and a way of making data is not at all rigid. Many researchers would speak of focus groups or participant observation as methods: They are ways of making data, with goals that fit these ways of making data, and each has a methods literature. But we prefer to consider these *strategies;* in Chapter 3, we discuss these as "incomplete" methods.

In the chapters to come, we discuss types of data, ways of handling data, and analytic techniques that belong to no particular method and are used in many. For now, our goal is to suggest the ways some data fit some methods. This does not mean that a way of making data is a method or implies a method. The fact that you are interviewing people tells an observer nothing about why, or about what you will do with those data. But the content and form of interviews and what you see in them will be different for different methods. This is because *how you think about the data* differs from method to method.

Table 2.2	The Fit of Method and the Type of Data
Chosen Method	*Likely Data Sources*
Phenomenology	*Primary:* audiotaped, in-depth conversations; phenomenological literature
	Secondary: poetry; art; films
Ethnography	*Primary:* participant observation; field notes; unstructured or structured interviews (sometimes audio- or videotaped)
	Secondary: documents, records; photographs; videotapes; maps, genograms, sociograms; focus groups
Grounded theory	*Primary:* interviews (usually audiotaped); participant and nonparticipant observations; conversations recorded in diaries and field notes
	Secondary: comparative instances; personal experience

From Choosing Sources and Sorts of Data to Managing and Analyzing Data

There is a further link in this methodological chain of research purpose, research question, choice of method, and the type of data needed. It is hardly surprising that the ways the researcher handles, manages, explores, and analyzes data are all part of the same chain.

Once again, our simple tabulation of the three methods shows commonalities. It is possible to describe in similar terms the strategies, or techniques of analysis, used within each of these different methods. The difference is not in the technique per se but in the way the strategy is used. Different ways of approaching the research will mean that the same techniques are used in different ways and produce different results. For example, researchers using very different methods may all code and, while coding, use the same technique—selecting a portion of text and assigning it to a category. But the similarity ends there. For each of them, the *way of approaching and thinking* about the data means that codes are applied in a particular way, and this results in *a particular way of linking data to categories.* The differences show when we ask questions such as the following: What is a category? What data are coded there? Is the collection of data for a category the end or the beginning of analysis? How do you think about the category, and how do you use categories? The answers are very

Table 2.3	The Fit of Method, Data, and Analysis Techniques	
Method	*Data Sources and Types*	*Analysis Techniques*
Phenomenology	Audiotaped, in-depth conversations; phenomenological literature	Theme-ing, phenomenological reflection Memoing and reflective writing
Ethnography	*Primary:* participant observation; field notes; structured or unstructured interviews *Secondary:* documents; focus groups	Thick description, rereading notes, storing information, and coding by topic; storying; case analysis Coding, recording field notes, and diagramming to show patterns and processes
Grounded theory	*Primary:* audiotaped interviews; observations *Secondary:* comparative instances; personal experience	Theoretical sensitivity, developing concepts, coding at categories, open coding for theory generation Focused memoing, diagramming, emphasis on search for core concepts and processes

different from method to method. Although different qualitative methods may utilize similar strategies, *how you think* while using particular strategies differs. We can expand Table 2.2, adding the mode of handling data and the analysis that fits; the results are displayed in Table 2.3.

☰ METHODOLOGICAL CONGRUENCE

In explaining the purposeful nature of qualitative inquiry, we arrive at our second principle of qualitative methods. Tables 2.1–2.3 show the way projects acquire *methodological congruence*—that is, fit between the research problem and the question, fit between the research question and the method, and, of course, fit among the method, the data, and the way of handling data. All of these components of the research process mesh to make the best possible end product. Thus each method is a distinctive way of approaching the world and data.

The concept of methodological congruence does not mean that data sources or analysis methods are predetermined for the researcher once he or she has chosen a method. It isn't that easy. Nor does it mean that a researcher has no flexibility once he or she has embarked on a particular path. *Methodological congruence* refers to the fact that projects entail congruent ways of thinking. The researcher working with phenomenology must learn to think phenomenologically if the fit of purpose, method, and data is to work well. If you are working with grounded theory, it is important that you learn how to think as a grounded theorist. The same sorts of data (e.g., field notes) will be interpreted differently by researchers using different methods, and similar data analysis techniques (e.g., coding) employed by researchers using different methods will have quite different analytic results, *because each researcher is thinking a different way.*

Qualitative research is not just a matter of performing techniques on data; rather, each qualitative method is a specific way of thinking about data and using techniques as tools to manipulate data to achieve a goal. Each component of the research process is linked to the next, and the chosen method dictates combinations of strategies to be used in particular ways to ensure consistency throughout the research process. As we show in Chapter 3, not all methods are as complete as the ones sketched in Tables 2.1–2.3. But all methods entail certain distinctive ways of thinking.

Seeing Congruence by Doing It

The webs of methodological congruence are most easily illustrated by an exploration of the different ways a real research topic can be handled. In what follows, we present a fictitious project concerning human attachment. If you have data from a previous study or a growing sense of your research interest, you might try applying what you read below to your approaches to that topic.

What is "human attachment"? Which literature should we look to? We have many choices—we could look at the literature on bonding between mothers and infants, at the family studies literature on family relationships, or even at the social support literature. We could extend this to the relationship literature on interaction, the literature on marriage, or the literature on mothering. We could choose a situation in which we could observe the concept as well as obtain personal accounts of attachment. From our broad topic and scan of the literature, let's choose to study public displays of attachment behavior at the arrivals and departures gates at airports. There we could observe attachment (and detachment) behaviors as passengers depart or as they greet family and friends on arrival. We could interview individuals (the passengers themselves or their relatives and friends) about the experience of greeting and leaving. Or we may consider interviewing "experts" who have observed many passengers greeting or leaving each other (porters, staff at car rental booths, security personnel waiting to check carry-on luggage, cleaning staff, and so on).

Given this topic (human attachment) and having identified a research context, our next step is to create a research question. Different questions will lead us to particular methods, and the method in turn will help us to decide details of the research design, such as who the participants will be, what the sample size should be, how data will be created and analyzed, and, most important, what type of results we will obtain.

Let us explore the topic by conducting an *armchair walkthrough*—that is, by taking a mindful stroll through the topic and visualizing what it might look like when we anticipate doing the study using each of the three major methods sketched above. The first concern of all qualitative researchers is locating the project. The setting for the research must be one in which the phenomena of interest are likely to be seen—frequently, and in an intense form. Those we choose to interview must be "expert participants," with much experience with the phenomena of interest. We must deliberately and purposefully select a setting or context where we will best see what we want to study. We do not usually choose a place or a sample randomly, for we would then have to rely on luck to see what we are interested in; we do not choose the "average" experience, as then the characteristics of the phenomena are diluted and less evident.

⟋⟋ THE ARMCHAIR WALKTHROUGH

How does one prepare to do a research study? Obviously, one may approach a particular problem in several different ways, developing several different questions, so that each one could be answered using a different method and could produce a slightly different result. Which one is best, and how is that determined?

One way to reduce the uncertainty is by conducting an armchair walkthrough—that is, by mentally going through the process. If I ask *this* research question, then I will need to use *this* particular method, seek *this* type of data and involve *these* participants, ask *these* interview questions, handle and analyze data *this* particular way, and the results will be in *this* form. On the other hand, if I do it using *that* method, then I will ask the questions *that* way, use *that* method, and involve *those* participants; data will look *that* way, and my results will be in *that* form.

By conducting an armchair walkthrough, we are trying to predict the research process and the outcome rather than go into research blindly. In this way, without losing flexibility or the ability to change some of our choices, we can focus on the data rather than on decisions about the administration of research. Although this type of conceptualizing will not detect every problem that may be encountered, it lets us get some sense of what we may learn by using each method. It allows for some level of informed choice about which method has the potential to provide the most suitable type of results, and it is helpful as we make preliminary preparation for writing the proposal. On the other hand, we need to be aware that such decisions are not carved in stone, and we should always be prepared to reevaluate and make changes if necessary. Table 2.4 displays the thinking that came out of the armchair walkthrough for our hypothetical project "Arrivals and Departures: Patterns of Human Attachment."

⪡ AND NOW—YOUR TOPIC?

"What are you studying?" is possibly the most common question asked of the researcher, and it is also quite often the most troublesome one. Interestingly, the issue of how to find a topic is not answered in any of the textbooks on qualitative research. This is because when you select a topic, you still have not started the research project. Selecting a topic involves also seeing the *purposiveness* of the study and the *congruence* of question, method, and what your project will be like.

Selecting the topic also involves selecting where you will go to do the study—it is not the research question you ask when you get there, or the method you use to answer it. If you find yourself telling inquirers, "I'm *doing* classroom authority/nurses' experiences of chosen childlessness/ inflicting pain . . . ." listen to the words you are using. The researcher does not "do" a topic as the mindless tourist "does" Belgium, checking off

Table 2.4 Comparison of Three Methods to Conduct a Hypothetical Project, "Arrivals and Departures: Patterns of Human Attachment"

Method	Research Question	Setting and Participants	Strategies	Types of Results
Phenomenology	What is the meaning of separation from or rejoining your spouse?	Interviews at interviewees' convenience; person who has traveled recently. 6–10 in each group	In-depth audiotaped conversations. Reflection on the phenomenological literature and other sources.	In-depth reflective description of the experience of separating from or rejoining your spouse.
Ethnography	What are the patterns of human attachment displayed during arrivals and departures at the airport?	Airport departure and lounge arrival - passengers, friends, relatives, experts at the scene (porters, airport personnel). Approximately 30–50 informants.	Unstructured, audiotaped interviews and participant observation at the gate. Field notes and other documents.	Description of the patterns of greeting behaviors or styles of farewell.
Grounded theory	What is the process of greeting or leaving your family?	Interviews anywhere, observations at the airport gate of passengers, family members. Approximately 30–50 participants.	Audiotaped interviews and observations. New data as theory directs research.	Theory about leaving and reunion; focus on the social psychological processes.

Source: Reprinted with permission from Sage Publications, Inc.

museums between France and Scandinavia. The *topic* of a research project is where it is *located*, where you are going to place your study, not what you will ask, how you will ask it, or how your research will provide answers when you are there. (Incidentally, the term comes from Aristotle's *Topics*, which contains common*place* arguments, from the Greek *topikos*, "of a place.")

A topic may be any researchable area, subject, or experience (such as an organization, living in a community, or having a particular learning disability), a concept (such as corporate structure, classroom learning, social support, or coping), a setting (such as a boardroom, a school, a village, a hospital ward), a group of persons (such as teachers, doctors, or teenagers), some aspect of their everyday activities (such as teachers' talk in the lounge), or activities that are unusual (teaching students with dyslexia). Those are all research locations or areas within which research questions can be defined. A topic may combine perspectives, so a researcher may be able to make an important argument for studying one of the above topics in a particular group by asserting that the experience of that group is sufficiently different from the experiences of other groups reported in the literature.

You may have several topics burning to be researched. The challenge, then, is to walk through each, asking how questions would be framed and what sorts of research they would require. Or you may have no topic, but instead a requirement that you get a project up and running. It seems harder to start that way, because then research presents itself as the push of duty, not the pull of interest in a topic. Wanted: a good topic!

How to Find a Topic

Any attempt to summarize reasons for selecting a topic runs the risk of appearing to present the process as orderly. It usually is not. Insights about suitable topics occur to researchers as they stand on high hills, while they are in the shower, or when they are in the library; topics demand attention when you are trying to do something else. A sort of typology is possible, however. If you are stumped, try locating your research in each of the five ways listed below. But remember to locate the project to ask how your topic would be studied and what the outcome project would be like.

You Are Already There

Statistically, "already being there" is undoubtedly the most common reason for topic selection. It is also the most exciting and the most dangerous. Because you are there, you possibly have, or may be convinced that you have, intimate knowledge of the topic as a participant. It seems you can get going fast—the preparatory work has been done. You are familiar with the setting and comfortable with the people there. But be careful: You were there for reasons other than research (such as employment or group membership or shared experience). These required a different type of preparatory work for you to become a good participant or actor in the setting. Being a researcher there may perhaps provide you with the opportunity to contribute new knowledge to an area you care about. And so you may, but you will have to ensure that your contribution represents valued research results and not merely what you wanted to prove or get done as a participant. If these ends are the same, you will have to be especially careful to establish that they were the same and that your study is rigorous.

There Is a Gap in the Literature

Topics that are amenable to qualitative inquiry have often been relatively ignored in the literature. Of course, this may be because they are inaccessible to researchers or, worse, simply uninteresting. The fact that nobody has studied a particular topic is not a good reason for taking it up. On the other hand, such topics may be neglected because they are areas in need of qualitative inquiry, areas where it is not easy to frame clear questions, areas that are difficult to access, or areas that are obscured by received interpretations.

Of course, this is a double-edged sword. If a topic has not been investigated, you will have an explorer's challenge of discovering a new place, mapping the area, displaying it to an admiring world, maybe even getting your name on it. Classic qualitative research projects have opened up whole areas of investigation in this way. With the second wave of feminism, qualitative studies returned to topics in the hitherto taken-for-granted social lives of women, opening up research areas addressing motherhood, social support networks, and even housework.

However, as Columbus found, undiscovered places are hard to sell. This is particularly important if you are a student applying for funding for research expenses. Research into topics that are "fashionable"—that is, topics that a number of other researchers are also investigating (or have investigated)—is generally easier to get funded, but there is usually a considerable amount of literature on those topics in the library already.

Another Way of Looking Is Needed

You might suspect that the literature may be poorly focused, or that there is something wrong, invalid, or inaccurate about the presentation and interpretation of the topic. Perhaps the received knowledge does not fit with the evidence, or results of the studies reported in the literature have been presented within the context of a theory that is invalid or inappropriate. It is time to take a fresh look at the phenomenon and reexamine the theory from within, taking into consideration the perceptions of those being studied. In recent decades, women's studies and studies of health and illness exemplify this approach, as qualitative studies challenged the functionalist paradigm, reopening questions of power and conflict.

What's Going On Here?

Qualitative methods are frequently used to discover the answers to quite pragmatic questions, such as "What is going on here?" or "How are we doing with this innovation?" Evaluation studies are of this type: The researchers are trying to understand and describe efficiently the processes or structures of particular phenomena. Much action research sets out to find out "what is going on here"—the topic is "here," this community, this fight, this local government organization, and so on.

Supplementing Quantitative Inquiry

The topic may be an area where there is considerable knowledge of events or patterns from quantitative research, or where quantitative work needs prior backgrounding. The qualitative project may form the groundwork for subsequent quantitative inquiry or be used to supplement quantitative inquiry, or quantitative inquiry may be used to illustrate qualitative inquiry. The end result of a qualitative project may be insight into a problem, a rich description, a hypothesis, a theory to be tested further in quantitative research, or a qualitatively derived theory that is ready to use. You should consider the purpose of the qualitative project before commencing the project and selecting the method.

From Topic to Researchable Question: Focusing Qualitative Inquiry

Deciding on a topic locates your research; this is where you are researching. Framing a qualitative question is harder, because it requires that you think about what needs to be asked in this research location as well as what you can ask and reasonably expect have answered given your resources and skills. A research question is a starting point only if it is researchable.

One of the most difficult tasks for the beginning researcher is to think qualitatively before the research begins. A researchable qualitative question is not the most obvious outcome of reflecting on a topic. The big first questions are as follows:

⦚ *What* needs to be asked?

⦚ *How* should it be asked? What data are required, and where will the researcher have to go to find answers to these questions?

⦚ *Can* it be asked, and if so, what sort of a researcher or research stance is needed?

Ethical as well as practical considerations must be explored (we discuss these in detail in Chapters 10 and 11). If you are planning to do research with vulnerable populations (such as groups in schools, prisons, hospitals, or some cultural groups), you must obtain special permission at the institutional level as well as at the guardian or parent, care provider, and individual levels. Once you have obtained access, you must have in place strategies to protect the identity of the participants. Consider who will have access to the raw data. How will it be stored? How will identities of participants or places be protected? Who will have access to the final report? And who will need to review it or approve it prior to publication?

⦚ WHAT CAN YOU AIM FOR?

By now it should be clear that qualitative researchers are aiming for an outcome that is more than a good story. It's the fit of method, data, and analysis that makes the difference between journalism and qualitative research. Good journalism and good qualitative research share goals of understanding people's situations, thoroughly researching and vividly illustrating what's found. But all qualitative methods aim for abstraction and analysis, not *only* description. (Robert Park, a founder of the Chicago School of Sociology, and a journalist by training, called sociology "journalism with a theory.")

And it will be a particular sort of analysis. In all the examples given above, the outcome is something *new*, a discovery *from the data*. This goal explains much in the techniques for handling data throughout this book. Qualitative coding, for example, aims to retain the detail of the data, so it can be explored and rethought. The researcher resists, or delays, reducing that detail to numbers, since to do so would prevent further discovery. Unlike much (though not all) quantitative research, the qualitative project is unlikely to be testing existing theories. Much more likely is that from the data will be created a new theory or a new explanation of the phenomenon studied.

These are not unreachable goals. Discovered theories may be very small and local. In Chapter 7, we discuss the task of abstraction and the ways it is done. Meanwhile, as you work toward a topic, ask, what could you aim for? What would be a good outcome of this study? What would be good enough, and what would be excellent? (For discussion of possible study outcomes, see Richards, 2005, pp. 125–145.)

⧽ ſUMMARY

We see the principles we have discussed in this chapter—the purposiveness of qualitative inquiry and methodological congruence—as the hallmarks of good qualitative research. They mean that a project's goals and its methods cannot be considered separately or severed from the strategies of a research design. A research strategy is only a *tool,* and how one uses a tool depends on the purpose of inquiry, the method used, and the type of data. This is important: One may learn a strategy, but *the way one uses it depends on the method.* In Chapter 3, we start to sketch this process.

In this chapter, we have emphasized the wholeness of methods—the fit of question, data, and analysis. In Chapter 3, we address the flip side of this wholeness: Although qualitative methods are congruent, they are not always complete, and they do not always fully direct each stage of the project. We compare the same three methods discussed above in terms of completeness, showing how some convey full instructions for the entire project whereas others leave the researcher to choose a methodological path.

REJOURCEJ

Read different types of qualitative research studies to get a feel for the differing results.

Morse, J. M. (Ed.). (1992b). *Qualitative health research*. Newbury Park, CA: Sage.
    This book provides a brief overview of the main types of qualitative inquiry and includes articles of each type as examples. It is useful for comparing and contrasting the types concerning the results that might be expected from research using the different methods.

Richards, L. (2005). *Handling qualitative data: A practical guide*. London: Sage.
    Chapters on each stage of tasks involved in doing the research, and on being able to see the project as a whole. Chapter 7 has a sketch of possible outcomes.

Other Resources

Brizuela, D., Stewart, J. P., Carrillo, R. G., & Garbey, J. (2000). *Acts of inquiry and qualitative research*. Cambridge, MA: Harvard Educational Review.
Creswell, J. W. (1998). *Qualitative inquiry and rsearch design: Choosing among five traditions*. Thousand Oaks, CA: Sage.
Eisner, E. W., & Peshkin, A. (Eds.). (1998). *Qualitative inquiry in education: The continuing debate*. New York: Teachers College Press.
Riessman, C. K. (Ed.). (1994). *Qualitative studies in social work research*. Thousand Oaks, CA: Sage.
Van Maanen, J., Dabbs, J. M., & Faulkner, R. R. (1982). *Varieties of qualitative research*. Beverly Hills, CA: Sage.

Qualitative Research by Discipline

We provide here a recent text in each of a range of disciplines, as a starting point for your reading in the relevant literature.

Daymon, C., Holloway, I., & Daymon, C. (2002). *Qualitative research methods and public relations & marketing communications*. London: Routledge.
Gilgun, J. F., Daly, K., & Handel, G. (Eds.). (1992). *Qualitative methods in family research*. Newbury Park, CA: Sage.
Golding, C. (2002). *Grounded theory: A practical guide for management, business, & market researchers*. Thousand Oaks, CA: Sage.
Holloway, I. (2005). *Qualitative research in health care*. Oxford, UK: Blackwell Science.
Latimer, J. (Ed.). (2003). *Advanced qualitative research for nursing*. Oxford, UK: Blackwell.
Mariampolski, H. (2001). *Qualitative market research: A comprehensive guide*. Thousand Oaks, CA: Sage.
Merriman, N. B. (1997). *Qualitative research and case study applications in education*. Toronto: John Wiley & Sons.

Munhall, P. L. (2001). *Nursing research: A qualitative perspective*. (3rd ed.). Boston: Jones & Bartlett.

Patton, M. Q. (2002). *Qualitative research & evaluation methods* (3rd ed.). Thousand Oaks, CA: Sage.

Shaw, I. S., & Gould, N. (2001). *Qualitative research and social work*. Thousand Oaks, CA: Sage.

Ulin, P. R., Robinson, E. T., & Tolley, E. E. (2005). *Qualitative methods in public health: A field guide for applied research*. San Francisco, CA: Jossey-Bass.

Journals

Ethnography

Field Methods

Forum: Qualitative Social Research (http: //qualitative-research.net/ fqs/fqs-eng.html)

International Journal of Qualitative Methods (http: //www.ualberta.ca/~ijqm)

International Journal of Qualitative Studies in Education

International Journal of Qualitative Studies on Health & Well-Being

Journal of Contemporary Ethnography

Qualitative Health Research

Qualitative Inquiry

Qualitative Report (http://www.nova.edu/ssss/qr/index.html)

Qualitative Research (http=//www.sagepub.com/journalsProdDesc.nav?prodId=Journal 201501)

Qualitative Research Journal (http//www.latrobe.edu.au/aqr/)

3

Selecting a Method

The key to doing qualitative research is selecting the best method to answer your research question. As in quantitative inquiry, there are many methods to choose from, and the underlying philosophies of the various methods provide distinct (and different) perspectives on reality. Although all methods may have certain procedures in common (such as coding) or may share some features with other methods (such as categorizing or "theme-ing"), each method has its own particular strategies and analytic techniques. The differences among methods give each method a unique perspective, and you must select the one that will enable you to achieve the goal of answering the research question.

⬈ COMMONALITIEƒ AND DIFFERENCEƒ

What Do All Qualitative Methods Have in Common?

All qualitative research seeks understanding of data that are complex and can be approached only in context. The methods we sketch in this book differ widely in how they do this and what the results look like, but some analytic processes are used in all qualitative research. Miles and Huberman (1994, p. 9) identify six analytic "moves" or strategies that are used in different ways within different methods. They are as follows:

- ⬈ *Meeting and coding* data as data records are created

- ⬈ *Recording reflections* and insights

- ⬈ *Sorting and sifting* through the data to identify similar phrases, relationships, patterns, themes, distinguishing features, and common sequences

※ *Seeking patterns or processes, commonalities and differences,* and extracting them for subsequent analysis

※ *Gradually elaborating a small set of generalizations* that cover the consistencies discerned in the database

※ *Confronting these generalizations* with a formalized body of knowledge in the form of constructs or theories

The Distinctiveness of Qualitative Methods

Although some analytic strategies may appear common to some methods, it is their application within each method that makes the various qualitative methods different from one another. The key to their differences is in the *way the researcher thinks about the data* and subsequently *conceptualizes*—that is, "thinks up" from data. This means that although similar techniques may be used, how different strategies are applied will produce different products in forms that fit the particular methods used and the questions being addressed. In this chapter, we briefly introduce three of the major qualitative methods. In later chapters, we address the generic processes of coding, categorizing, and theme-ing and reintroduce the strategies that make methods distinct from each other.

※ PHENOMENOLOGY

Phenomenology is one of the most important philosophical movements of the twentieth century. Founded by Edmund Husserl (1859–1938), it is used to refer to both a philosophy and a research approach. As a method, it has undergone many shifts, orientations, or approaches. Van Manen (2006) has classified the following:

※ **Transcendental phenomenology:** (Husserl and his collaborators: Eugen Fink, Tymieniecka, Van Breda, and Giorgi) These phenomenologists present phenomenology as an interpretative, rather than an objective, mode of description. This interpretation is presuppositionless and based on "intentionality" ("all conscious awarenesses are intentional awarenesses") and "eidetic reduction" (vivid and detailed attentiveness to description). Transcendental phenomenology explores the way knowledge comes into being, and knowledge is based on insights rather than objective characteristics, which "constitutes meaning."

Selecting a Method 49

✒ **Existential phenomenology:** (Heidegger, Sartre, de Beauvoir, Merleau-Ponty, Marcel, and others) believe that the observer cannot separate him-/herself from the lived world. "Being-in-the-world" is reality as it is perceived, and there is a reciprocal relationship between the observer and the phenomenon that includes all thoughts, moods, efforts, and actions within the life-world that is man situated. Pre-reflected experiences, the life-world, and phenomena constitute *existence,* or human reality.

✒ **Hermeneutical phenomenology:** (Heidegger, Gadamer, Ricoeur, van Manen) believe knowledge comes into being through language and understanding. Understanding and interpretation are intertwined, and interpretation is an evolving process. Hermeneutic phenomenologists use culture (symbols, myth, religion, art, and language), poetry, and art in their interpretations.

✒ **Linguistical phenomenology:** (Blanchot, Derrida, Foucault) take the perspective that language and discourse reveal the relations between "understanding, culture, historicality, identity, and human life." Meaning "resides in language and the text, rather than in the subject, in consciousness, or even in lived experience."

Here we describe the *hermeneutical phenomenology,* as a method. From this perspective, phenomenology offers a descriptive, reflective, interpretative, and engaging mode of inquiry from which the essence of an experience may be elicited. Experience is considered to be an individual's perceptions of his or her presence in the world at the moment when things, truths, or values are constituted (van Manen, 1990).

Four existentials guide phenomenological reflection: *temporality* (lived time), *spatiality* (lived space), *corporeality* (lived body), and *relationality* or *communality* (lived human relation) (van Manen, 1990). People are considered to be tied to their worlds—embodied—and are understandable only in their contexts. Existence in this sense is meaningful (*being in the world*), and the focus is on the lived experience. Human behavior occurs in the context of relationships to things, people, events, and situations.

Assumptions

Two major assumptions underlie phenomenology. The first is that perceptions present us with evidence of the world—not as it is thought to be, but as it is lived. The lived world, or the *lived experience,* is critical to phenomenology. The second assumption is that human existence is meaningful and of interest in the sense that we are always conscious of something. Existence as *being in the world* is a phenomenological phrase acknowledging that people are in their worlds and are understandable only in their contexts. Human behavior occurs in the context of the four existentialisms introduced above: relationships to things, people, events, and situations.

What Sorts of Questions Are Addressed?

Phenomenological inquiry may or may not be formalized as a question per se. The researcher may have an interest simply targeted toward understanding the meaning of the lived experience in a particular phenomenon, with questions arising as inquiry proceeds. Therefore it is these questions that sensitize inquiry in the study. For instance, in considering his experiences as a parent of a child undergoing a heart transplant, Smith (1989/1992) describes his frustration with the delays in obtaining postoperative analgesic for his child:

> The interviewer does not simply ask a question of whose interests are being served—the parents' or the child's? But rather, he asks, how can a medical decision be made in presumably the best interests of the child by ignoring those of us who have been responsible by now for the welfare of the child? If we, the parents of a particular child, want to remain close to our child, what might we be up against when a crucial medical decision is made as to what should be done for our child? What sort of logic would deny fundamental responsibility we feel for our child? (p. 106)

In published reports, the research questions are often embedded in the introductory remarks that set the context of the study. For instance, Kelpin (1984/1992), in her study of birthing pain, notes that the pain of childbirth has a particular "centrality" for women's relationships as mothers and as human beings. The way Kelpin considers her research topic provides us with an excellent example of the way phenomenologists consider their research questions:

> What do the pains of birth tell us about ourselves, about our sufferings and our joys? Is there something in the pangs of childbirth which holds true for all women: those who pleasure and ride above the

pain? Those who endure it? And those who suffer? Some birthings are short and intense, some are long and exhausting, and some in need of medical intervention and treatment with forceps, medication and Caesarian delivery. Is it possible that viewing pain-as-lived may reveal sublimity and joy as well as the agony, the hurtfulness of the pain of childbirth? Our immediate appraisal of pain-as-experienced may bring light to inner meanings that go beyond theoretical and practical approaches. By coming to an understanding of the pain as experienced by women we may be able to come to grips with the significance or essence of the pain. (pp. 93–94)

Van Manen (1990) notes that stating a question directly often simplifies the problem, so in phenomenology, the actual research question may be left implicit. Clarke (1990/1992), for instance, explores her child's experience of asthma in light of her own reflections on her child's experience. She does not state the question explicitly but introduces it in the phenomenological way, using voices of her daughter's essay ("Memories of Breathing"), voices of poets as illustrators, and voices from the phenomenological literature, while her own voice guides our insights into the experience.

Researcher Stance

When thinking phenomenologically, the researcher attempts to understand, or grasp, the essence of how people attend to the world (using the four existentialisms), remembering that a person's description is a *perception,* a form of interpretation (Boyd, 1993; van Manen, 1990). Every day, we consciously experience concrete objects through intuition. Giorgi (1997), on the other hand, notes that *presences* are the experience of many phenomena that are not "realistic" but are vital to the understanding of the lived experience. These are such things as dreams and delusions. *Intentionality* is the essential feature of consciousness. Consciousness is always "directed to an object that is not in itself conscious, although it could be, as in reflected acts" (Giorgi, 1997, p. 236).

Data Gathering/Making

Phenomenological researchers *bracket* all a priori knowledge about the topic; by writing their assumptions, knowledge, and expectations, they enter the conversation with no presuppositions. They most frequently

gather new data by using audiotaped "conversations" without predetermined questions, following a "clue-and-clue-taking process" as the conversations proceed (Ray, 1994, p. 129). They then transcribe these taped
conversations and use them as a basis for reflection. During analysis,
phenomenologists also reflect on personal experiences, observations,
and the experiences of others—even those expressed in poetry, literature,
and film.

What Do the Results Look Like?

Phenomenology gives us insights into the meanings or the essences of
experiences that we may previously have been unaware of but can recognize. This experience of confirmation is known as the *phenomenological
nod.* The essence may be presented in an essay as several segments or
perspectives, each describing a different dimension of the experience.
Phenomenological researchers may share the results of their studies in
essays or in book-length works.

What Forms May Phenomenology Take?

As discussed above, phenomenological methods have evolved in more
than one direction and take several forms that have some commonalities.
All phenomenologists subscribe to the belief that being human is a
unique way of being, in that human experiences and actions follow from
their self-interpretation (Benner, 1994b, p. ix).

Van Manen (1990) refers to his method as a *hermeneutic phenomenological approach;* it starts with the exploration of a pedagogically grounded
concept within the everyday lived experience. Through processes of reflection, writing and rewriting, and thematic analysis, the researcher may
describe and interpret the essence or meaning of the lived experience.

Giorgi (1997) uses the term *phenomenology* in the descriptive Husserlian
sense. His methods consist of three interlocking processes: *phenomenological reduction, description,* and a *search for essences.* We address these
processes in more detail in Chapter 8.

Researchers use *heuristic phenomenology* (Moustakas, 1994) when
they seek to understand themselves and their lived worlds. Although
such research is autobiographical, the questions it answers may have
social, and even universal, significance. Heuristic research "unfolds"

through initial engagement, immersion into the topic and the question, incubation, explication, and culmination of the research in "creative synthesis" (Moustakas, 1990).

⧥ ETHNOGRAPHY

Ethnography provides a means for exploring cultural groups. But what is culture? Culture is an abstract concept used to account for the beliefs, values, and behaviors of cohesive groups of people. It is a narrower term than *race* (which accounts for biological variation); a racial group may contain many different cultures, and a cultural group may contain members of different races. Although *cultural group* may refer to a particular nationality, cultures may cross political boundaries and a nation may contain many cultural groups. Culture is responsive to change, and people adapt to new cultures—for example, when two groups come into contact and their members intermarry, or when families from one culture migrate into territory occupied by persons of another culture, as occurs with immigration and other forms of relocation. However, at a broad level, culture accounts for many perceptions and the ways in which people view their worlds.

Thus, although there are hundreds of definitions of culture, most include the notion that culture consists of the "beliefs, behaviors, norms, attitudes, social arrangements, and forms of expression that form describable patterns in the lives of members of a community or institution" (LeCompte & Schensul, 1999a, p. 21). Within a cultural group, behaviors are patterned and values and meanings are shared, and these patterns vary from culture to culture. Ethnography is always holistic, contextual, reflexive, and presented from the emic perspective (Boyle, 1994).

Assumptions

Because cultural assumptions, beliefs, and behaviors are embedded within a cultural group, they are not always evident to those who are a part of the group. Thus ethnography is best conducted by researchers who are not part of the cultural group (i.e., from the *etic* perspective) and is facilitated by researchers' comparing and contrasting two cultures. If a researcher shares the participants' culture (as in the *emic*, or insider, perspective), then it is difficult for him or her to "see" the beliefs, values, practices, and behaviors embedded in everyday life. The research will be easier, and the differences more evident, if the researcher is an outsider to both of the cultures being compared and contrasted. Theories of culture generally agree that cultural beliefs, values, and behaviors are learned and transmitted within the group, and that this learned worldview is transmitted and shared among individuals. This does not mean that culture is static—on the contrary, it is *dynamic, adapting,* and *changing.* Overall, however, cultural beliefs and practices are *patterned,* and the strategies used within ethnographic methods are designed to elicit the features *implicit* in a culture.

Based on the assumption of culture as "shared values, beliefs, and behaviors within a cohesive group," researchers are now using ethnographic methods to explore smaller *subcultural* units, such as institutions, in particular closed institutions (e.g., prisons, hospitals, and nursing homes), and to study loosely connected groups of people (e.g., hockey teams or motorcycle gangs), those with particular occupations (e.g., university professors or physicians), and persons with particular characteristics, such as a shared illness or condition (e.g., stroke patients or nursing mothers).

What Sorts of Questions Are Addressed?

Ethnographic research explores phenomena within cultural contexts from the emic perspective, or from the perspectives of the members of the cultural groups involved. Davis (1986/1992), for instance, describes the meaning menopause holds for women of Grey Rock Harbour, "taking into account both the collective and idiosyncratic elements of village life which help explicate the emic perspective of menopause. These include (1) the semantics of menopause, (2) lay semantics and (3) local institutions and the moral order" (p. 151).

Cassell (1987/1992) used ethnographic methods to explore the work of surgeons as well as the ethos and "the set of traits distinctive of that profession." She also examined some of "the dynamics and the personal cost and benefits of maintaining the ethos and the set of traits" (pp. 170–171).

Researcher Stance

Ethnography is always conducted in the natural setting, or the *field,* so that the researcher studies the lives of members of the cultural group directly. Ethnographers work to become as integrated as possible into the lives of the people they are studying. Recall that the researcher stance is outside the group being studied, yet the data collection procedures are designed to elicit emic data (i.e., data reflecting the "native" point of view). The researcher is a student of the cultural group he or she is studying, learning and being taught yet not truly one of the group.

When an ethnographic researcher is collecting data, the fact that culture is shared among all group members theoretically means that any member of the group may serve as a participant in the study; however, the researcher must consider the characteristics of *good informants* (i.e., having the ability to reflect on and describe the culture, being articulate and patient) and the type of data required. Participants who work most closely with, and interpret the culture for, the researcher are known as *key informants*. They serve to inform and instruct the researcher about the culture, although the researcher compiles these data and verifies them with data from other participants. During data collection and analysis, the researcher must consistently reflect on the results in the context of the cultural values, beliefs, and behaviors of the group being studied.

Data Gathering/Making

Ethnographic data collection consists of strategies for obtaining data that will enable the researcher to describe cultural norms, perspectives, characteristics, and patterns. The research purpose and question dictate the types and forms of data collected. In ethnography, data are not usually of a single type; they consist of *observational data* (recorded as *field notes* or in the form of photographs, videotape, and so forth), *interviews* (recorded as *field notes* or audiotaped and transcribed), the researcher's ongoing theoretical notes in a *diary*, plus any other data that may be relevant. In turn, these data may take various *forms*. For instance, interviews may be unstructured, semistructured, or structured; they may include questionnaires, surveys, or special techniques such as sentence frames and card sorts to elicit particular kinds of responses that fall within the parameters of whatever is being studied. Quantitative data may also be included.

An ethnography usually has distinct *stages* and *phases*, during which different types of data are collected and the researcher's effectiveness as an analyst varies (Morse & Field, 1995, pp. 71–73). The *first phase* is "getting in," during which the researcher is a stranger to the setting and the primary task is *negotiating entry.* In this stage, the researcher goes through a process of finding a role and fitting in. The researcher feels awkward and self-conscious. Wax (1971) notes that one should not become an ethnographer unless one can tolerate feeling "out of place" and "making a fool of one's self" (p. 370). Usually, during this phase the researcher does not understand the setting or the participants, so interpretation is premature. Thus data making at this stage should focus on relatively concrete tasks, such as making maps of the setting or becoming acquainted with who's who in the community being studied. The researcher keeps a diary of initial impressions and uses field notes to record observations.

During the *second phase,* the researcher becomes better acquainted with the routines in the setting, and the participants become more comfortable with the researcher. Data making now consists of nonparticipant observations and informal conversations. The researcher may identify a key informant (participant), select initial participants, and begin interviewing. The researcher also commences analysis as he or she becomes more comfortable in the setting. The researcher may also develop some initial analytic hunches about the setting.

Trust has developed between the participants and the researcher by the time of the *third phase,* which is marked by *cooperation and acceptance.* Data making is most productive in this phase. The researcher now understands what is happening in the setting, and the data become more focused; the researcher also uses the data to verify hunches and to develop theoretical formulations.

At the end of the third phase, the researcher may feel relaxed and integrated into the setting, to the extent of becoming acculturated. A potential problem at this stage is that the researcher may identify more with the cultural norms of the group than with his or her own research agenda, and may lose objectivity in conducting observations and analysis. The *fourth phase* is therefore one of *withdrawal.* The research focus at this stage is primarily on data analysis, and data collection consists of gathering data to resolve ambiguities, to fill in areas that are thin, and to verify previous data. The task of the *last phase* is analysis; the research is completed and the ethnography written.

Awareness of self during data collection is vital. The key to good ethnography is the researcher's awareness of his or her own cultural values, beliefs, and biases and the way they influence what data are collected. The researcher must also be aware of roles and relationships with others in the field, what data are collected, and why they are collected. The researcher must record all these self-observations in a research diary, a document that can have a profound impact on the analytic pathway—how the researcher moves through the process of making sense of the data. It enables the researcher to see or not see the obvious and the less obvious. Relationships established between the researcher and those in the field, the development of trust, and the degree of the researcher's inclusion as a member of the group—all of these factors have some influence on the type and quality of data that are collected and available for analysis.

What Do the Results Look Like?

The main goal of most ethnographic research is what has been classically termed *thick description* (Geertz, 1973): a narrative that describes richly and in great detail all features of the culture. Other ways of presenting data include *taxonomies* or *classification systems*, whose function is to name objects and display the relationships among them, thus creating a framework for displaying the data. A taxonomy permits the researcher to display classes of objects according to common characteristics as well as the subcategories of related objects within a particular class. A taxonomy does not, however, account for processes.

Ethnography, perhaps more than any of the other methods sketched here, has undergone major changes in recent years. You will find that the literature contains a lively debate about the rival goals of description and theorizing. The results may look like theoretical monographs or like documentary films and articles on some aspect of daily life (e.g., eating, dance, health beliefs), special circumstances (e.g., childbirth, funeral ceremonies), or representation (e.g., use of media such as art, drama, dance).

What Forms May Ethnography Take?

Ethnography may take several forms, depending on the type of research question, its scope, and the researcher's perspective.

Traditional ethnography is conducted in a culture unfamiliar to the researcher and may focus on a single setting (usually a village or several villages). The research focus is broad, with the ethnographer describing the group as comprehensively as possible, including language, kinship systems, and so forth. This type of ethnography requires prolonged residence and engagement with the cultural group (of at least a year) and, if the goal is to document change, subsequent visits over a number of years.

Focused ethnography is used primarily to evaluate or to elicit information on a special topic or shared experience. It differs from traditional ethnography in that, as Muecke (1994) notes, the topic is specific and may be identified before the researcher commences the study. Focused ethnography might be conducted with a subcultural group rather than with a cultural group completely different from that of the researcher. It may also be used to study institutions. For example, Gubrium (1975) studied a nursing home, and Germain (1979) looked at a cancer ward. Ethnographic studies may focus on groups of participants who share some feature or features, such as a particular disability. In such studies, participants may not know each other, but the researcher focuses on their common behaviors and experiences resulting from their shared features, such as being treated the same way by care providers. This enables the researcher to apply the assumptions from a shared culture.

In focused ethnography, data making may include only some of the strategies that define ethnography. For instance, fieldwork may be omitted, and data may consist only of interviews.

Whereas the basic tenets of traditional ethnography place the researcher in an ambiguous position by necessitating his or her being both a part of the group and a distant observer, *autoethnography* legitimates the researcher's use of his or her own experience. In autoethnography, researchers analyze personal narratives in the light of sociological literature (see Ellis & Bochner, 1996, 2000). *Critical ethnography* examines cultural knowledge and action with the aim of forcing society to identify and act on values and ethical and political issues. Thomas (1993) notes that "critical ethnographers describe, analyze, and open to scrutiny otherwise hidden agendas, power centers, and assumptions that inhibit, repress, and constrain. Critical scholarship requires that commonsense assumptions be questioned" (pp. 2–3).

Participatory action research (PAR) follows the ethnographic methods of conducting field research using strategies of interviews and observations, but it challenges the researcher-participant relationship of ethnographic research. Rather than conducting research *on* people, practitioners of PAR conduct research *with* the people who are being studied (Reason, 1988). They believe that such *cooperative inquiry* is less likely to "undermine the self-determination of their participants" (p. 4). Participants discuss and agree on what they want to research, the nature of the questions, modes of data collection and analysis, the way data are written up, and how the findings are distributed.

Action research (AR) is research also conducted by a team of professional action researchers and stakeholders—members of the organization or community being studied—with the goal of *seeking to improve their situation*. As in PAR, they jointly define the research problem, cogenerate relevant background knowledge, identify and learn research methods, and interpret and implement the findings. Thus AR "democratizes the relationship between the professional researchers and the local interested parties" (Greenwood & Levin, 1998).

Since the 1930s, when Bates and Mead (1942) used film to document fieldwork, film, and now primarily video, has been an important medium for ethnography. In *visual ethnography*, video or film is used to record the scene, the daily lives of participants, interviews, and those events that cannot accurately (or with detail) be stored as field notes. Researchers may use videotaping or filming in two ways: to record, catalog, and gather data to supplement participant observation or as a stand-alone strategy for interpretation (for example, researchers may manipulate videotaped data, slowing down or speeding up the tape, to explore interactions or nonverbal gestures in microanalytic detail). In addition, video enables researchers to examine dialogue, along with its accompanying gestures, in detail (see Goldman-Segall, 1998).

〰 GROUNDED THEORY

Assumptions

Grounded theory has its origins in symbolic interactionism, taking the perspective that reality is negotiated between people, always changing, and constantly evolving (Blumer, 1969/1986). Research questions in grounded theory reflect this interest in process and change over time, and the methods of making and analyzing data reflect a commitment to understanding the ways in which reality is socially constructed. It is these processes of change and social construction that the researcher examines, identifying stages and phases. The assumption is that through detailed exploration, with theoretical sensitivity, the researcher can construct theory *grounded in* data.

The method was developed originally by Glaser and Strauss (1967) with equal attribution; as we show below, the idea of theoretical sensitivity

(Glaser, 1978) and the techniques for creating theory grounded in data (Strauss, 1987; Strauss & Corbin, 1990) were developed separately by the two authors. In their works, Glaser and Strauss presented researchers with what at the time was a radical proposal—that theory should be developed "in intimate relationship with data, with researchers fully aware of themselves as instruments for developing that grounded theory" (Strauss, 1987, p. 6).

Such theory will usually be small scale and focused, and techniques will emphasize the "continuous interplay between analysis and data collection" (Strauss & Corbin, 1994, p. 273) until a theory fitting the data is created. The process involves a data-driven design (theoretical sampling). The key goal is the creation of new theoretical concepts *from* the data and the seeking of *core concepts* (Strauss, 1987), or the pursuit of what Glaser (1978) terms the *basic social process* (BSP) or the *basic social psychological process* (BSPP).

What Sorts of Questions Are Addressed?

Grounded theory studies usually begin with questions about "What's going on here?" This is an appropriate method for the researcher wishing to learn *from* the participants how to understand a process or a situation. The questions themselves suggest the examination of a process. Thus grounded theory studies are usually situated in experiences in which change is expected, and the method has become dominant in research areas where the understanding of change and process is central, such as in health and business studies. For instance, Lorencz (1988/1992) explored the experiences of predischarge schizophrenics with at least 2 years of illness. Morse and Bottorff (1988/1992) studied mothers who were breastfeeding; in particular, they were interested in breast milk expression. They asked: "What are the perceptions of mothers who express? What are the feelings and attitudes that are evoked by expressing? How do mothers learn to express?" (pp. 320–321). Turner (1994) gives a detailed account of his techniques for discovery of a grounded theory in his description of the organizational processes that led to a ferry disaster.

Researcher Stance

The concept of *theoretical sensitivity* is crucial in grounded theory. The researcher *seeks* theory, constantly works with data records and records of ideas to tease from them the concepts and the linkages that might generate theoretical insight. Those emerging concepts are also in constant interplay with the data as the researcher seeks integration and synthesis.

The perspective that reality is constantly changing and being negotiated leads the researcher to active inquiry into the event over time. There is an emphasis on detailed knowledge, constant comparison, and the trajectory of the event. The researcher consistently asks not only "What is going on here?" but "How is it different?" The method of grounded theory promotes a stance of refusal to accept a report at face value, a sort of methodological restlessness that leads the researcher to seek characteristics, conditions, causes, antecedents, and consequences of events or responses as ways of drawing them together in an integrated theory.

Data Gathering/Making

Grounded theory research does not require any particular data source, but it does require data within which theory can be grounded. The goal of discovering theory from data sets high standards for the data, both in depth of detail and in coverage of process. However the data are made, the records must support the probing and friction of constant comparison and reflection. A study may commence with an observational phase in the field or with interviews—narratives about the event, told sequentially from beginning to end. Such interviews are much more able to support the method than are semistructured interviews or brief accounts.

Researchers should beware of attempting grounded theory research with structured data records, which preemptively limit what they will hear in response to their preconceived questions. In such data, it is difficult to identify the process or discover categories derived from the meanings held by others.

What Do the Results Look Like?

Grounded theory is undoubtedly the label most popularly applied to qualitative research, and undoubtedly the most misapplied, often being taken as synonymous with *qualitative* (Lee & Fielding, 1996). We share a concern that researchers should understand the true nature of grounded theory; it is a unique and highly demanding method, with strong congruence. If you don't know what method you are using, it is highly unlikely to be grounded theory.

The explicit goal of grounded theory studies is to develop theory—theory derived from, and grounded in, the data. A study using grounded theory will usually have a single story line, offering a core concept and its attendant theory as a way of making sense of the data. These are *new* theoretical offerings, not seen before that particular study, because they are the product of it. Often the core concepts are also new, not everyday, concepts.

A grounded theory study is densely argued; the researcher identifies the concepts involved and develops theory by exploring the relationships between these concepts in the stages or phases of the process and the core category or variable (or *basic social process*). This one category is the theme that runs through the data and accounts for most of the variance. A grounded theory study attempts to account for the centrality of the core concept by telling the story of its emergence. Reports may include diagrams of the process, or summary typologies, indicating the presence or absence of selected factors.

What Forms May Grounded Theory Take?

The founders of grounded theory came from contrasting backgrounds and worked as coinvestigators on high-profile projects. For two decades, based on the original work of Glaser and Strauss (1967), grounded theory was presented as a coherent and complete method—but as one method. Over the next two decades, as each author worked independently the method evolved and diverged, with Glaser (1978) and Strauss (1987) separately writing significant and very different methodological texts. At that time, most researchers using the techniques assumed there was a single set of methodological procedures for grounded theory research.

Divisions between Glaser and Strauss appeared in the early 1990s with a publication by Glaser (1992) in which he rejected Strauss's book coauthored with Corbin (1990). Instead of generating a methodological debate, this created in some locations two "schools" of grounded theory, termed *Glaserian* and *Straussian* grounded theory (Stern, 1994, p. 219). Researchers in these areas are increasingly (and arguably, regrettably) expected to choose between two distinct sets of procedures, using as their primary source either Glaser's (1978) or Strauss's (1987) text, further developed in Strauss and Corbin (1994, 1998), published after the death of Strauss. All methods evolve, and should do so. We share a hope that the various strands of grounded theory will be developed productively, to fit projects as appropriate, and we encourage you to explore and if appropriate draw on both these groups of techniques. But you should be aware that the following distinctions are commonly made.

Glaserian Grounded Theory

Glaserian grounded theory takes the more objectivist perspective: Data are both separate and distant from both the participants and the analyst (Charmaz, 2006). Glaser focuses his attention *on* the data to allow the data to tell their own story. The Glaserian analyst attends to the data and asks, "What do we have here?" (Stern, 1994, p. 220). As in the original documents on grounded theory, analysis focuses on components of the theory—on the processes, categories, dimensions, and properties—and it is the development of, and the interaction between, these components that allows the theory to emerge. In Glaserian approaches, the theory is more often diagrammed to illustrate the relationships between concepts and categories.

Straussian Grounded Theory

The so-called Straussian grounded theorist examines the data and stops at each word or phrase to ask, "What if?" Thus, the analyst "brings to bear every possible contingency that *could* relate to the data, whether it appears in the data or not" (Stern, 1994, p. 220). Straussian grounded theorists are concerned with striving to rise above the data to develop more abstract concepts and their descriptions. Theories are created in interaction with the data and (as in Glaserian approaches) retain the emphasis on categories, dimensions, and properties. There is a strong emphasis on "open coding," best exemplified in the taped research conversations in Strauss (1987). Theories are the product of reflection, discussion, and detailed examination of text, constructed from memos and dense coding. Straussian researchers rely less on diagrams than Glaserian grounded theorists.

Dimensional Analysis

A third early form of grounded theory (and one that is very different from the other two) is dimensional analysis, developed by Schatzman (1991), a colleague of Glaser and Strauss. Dimensional analysis allows for the "explicit articulation of the analytic process" and provides "an overarching structure to guide analysis" (Kools, McCarthy, Durham, & Robrecht, 1996, p. 314).

Constructivist Grounded Theory

In contrast to the Glaserian and Strauss and Corbin objectivist, constructivist grounded theory is interpretative—both the data and the analysis are

> created from shared experiences and relationships with participants. It provides a means to learn "how, when, and to what extent the studied experience is embedded in a larger and often, hidden positions, networks, situations, and relationships. Subsequently, differences and distinctions between people become visible . . . A constructivist approach means being alert to conditions under which such differences and distinctions arise and are maintained. (Charmaz, 2006, pp. 130–131)

Situational Analysis

Recently developed by Adele Clarke (2005), situational analysis focuses on the situation—context and people, their relations, actions, and interactions. It uses interview, observational, and other sources. Situational analysis "allows researchers to draw together studies of discourse and agency, action and structure, image, text and context, history and the present moment—to analyze complex situations of inquiry broadly conceived" (Clarke, 2005, p. xxii). Thus it differs dramatically from process-oriented grounded theory, in that the theory is not constructed around a basic social process. Rather, it is organized by a situation-centered framework developed by Anselm Strauss, using three types of mapping data, to organize "key elements, materialities, discourses, structures, and conditions that characterize the situation of the inquiry" (Clarke, 2005, p. xxii). In this way, "the *situation becomes the unit of analysis*, and understanding its elements and their relations is the primary goal" (Clarke, 2005, p. xxii, italics in the original). More closely aligned with ethnography than is traditional grounded theory, it enables the analysis of "highly complex situations of actions and positionality, of heterogeneous discourses . . . and situated knowledges and positionality, of the heterogenous discourses . . . and of the situated knowledges of life itself" (Clarke, 2005, p. xxiii).

The three types of mapping inherent in situational analysis described by Clarke (2005) are:

1. *Situational maps* delineating "major human, nonhuman, discursive, and other elements in the research situation of inquiry and [that] provoke analysis of relations among them.

2. *Social networks/arenas maps* that "lay out the collective actors, key nonhuman elements, and areas of commitment and discourse within which they are engaged in ongoing negotiates—meso-level interpretations of the situation; and

3. *Positional maps* that lay out the major positions taken, and not taken, in the data, vis-à-vis particular axes of difference, concern, and controversy around issues in the situation of inquiry. (Clarke, 2005, p. xxii)

We urge you to discover the differences in these approaches and evaluate their significance, avoiding the abyss created by claims that there is only one way to achieve grounded theory. Such claims ossify methods and prevent researchers from modifying recommended procedures or developing new ways of combining them. Our advice is that you return to the earlier works of the founders and that you note the tone of those writings. Strauss wrote in 1987 that the methods he describes "are by no means to be regarded as hard and fixed rules for converting data into effective theory" (p. 7). Rigid rules, after all, are particularly inimical to grounded theory approaches.

> Researchers need to be alive not only to the constraints and challenges of research settings and research aims, but to the nature of their data. They must also be alert to the temporal aspects or phasing of their researches, the open-ended character of the "best research" in any discipline, the immense significance of their own experiences as researchers and the local contexts in which the researches are conducted. (Strauss, 1987, pp. 7–8)

※ ADDITIONAL QUALITATIVE METHODS

We have illustrated our point about methodological congruence with sketches of only three major methods. The appropriate method for your study may not be one of these. Working from your research question, you may be led to other methods, such as narrative inquiry, conversation analysis, discourse analysis, hermeneutics, ethnomethodology, life history, or ethology. Use the resources listed at the end of this chapter to explore the range of methods—their assumptions, the sorts of questions they address, the researcher stances associated with them, the ways data are made, and what the results look like. Methods are tools. Research problems should not be modified to fit the researcher's repertoire of methods; rather, the researcher's goal should be to become versatile enough to identify and use the appropriate method for the research question.

Working this way, from question to method, you will not be tempted to approach qualitative research as though it were done by one generic method. A researcher who does so can get far into a study, even through it, with descriptive rather than analytic results. Identifying a question, conducting interviews, and then analyzing data by identifying themes or categories locate patterns but rarely produce a theoretical outcome. "Sorted data" as an end result are only as interesting as the data themselves. Such a study may be very interesting, but if conducted without the benefit of a coherent method, it will usually end at this descriptive level.

You will encounter many examples of such work. Our aim is not to condemn descriptive work, but rather to show how it differs crucially from research within a congruent qualitative method and why such descriptive work is often not regarded as qualitative or accepted for publication in qualitative journals (Morse, 1996). Concerned with the problems of researchers' attempting to retrofit a congruent qualitative method to data-sorting and pattern-finding tasks, Richards (2000) has labeled such descriptive work "pattern analysis." In short-term pragmatic studies especially, researchers may have no goals beyond seeking and reporting patterns in data, for example, by demographic variables such as gender or structural factors such as socioeconomic settings of schools studied. If the task is to find out whether the responses to an idea addressed in focus groups vary by gender, or to establish whether the level of acceptance of an initiative in schools is different in lower- and upper-class areas, unstructured data may be necessary and relevant, and analysis may not require abstraction. Often such studies combine qualitative and quantitative data skillfully and usefully, with the discovery of patterns in the unstructured data illuminating the statistical analysis. If your goal is pattern analysis, many of the techniques discussed in this book may assist you in discovering and reporting patterns. But don't represent your study as grounded theory.

It is the lens provided by a method that enables abstraction from data, the emergence and construction of theory about the data, and the linking of the results to the literature and other theories.

∭ ſUMMARY

In this chapter, we have tried to convey that the most significant feature of using a method is the researcher's perspective. Remember, to be a phenomenologist, you must think like a phenomenologist; to be an ethnographer, you must think like an ethnographer; to be a grounded theorist, you must think like a grounded theorist; and so forth. This is so important that Morse says she has "different tracks" in her brain for thinking in the various ways demanded by individual methods. Nevertheless, within the perspective of each method, the researcher manipulates data by using *analytic techniques*. Although these analytic techniques appear similar for all methods, *how they are used with the data* is what makes a method a particular method. Different methods may use similar techniques, but the individual method's strategy (the way the techniques are used) gives it a unique application and produces a unique result.

The goal of all qualitative inquiry is not to reproduce reality descriptively but to add insight and understanding and to create theory that provides explanation and even prediction. The best way to gain an appreciation for these differences is to read completed studies that provide examples of the various methods. Ask yourself: How do these studies differ? What contributions do each offer? What level of abstraction or theory development has each reached? Can you begin to identify how each of the authors has obtained abstraction and has, at the same time, come to understand the phenomenon in context?

∭ REſOURCEſ

Phenomenology: Methodological Resources

Benner, P. (Ed.). (1994a). *Interpretive phenomenology: Embodiment, caring, and ethics in health and illness.* Thousand Oaks, CA: Sage.

Boyd, C. O. (1993). Phenomenology: The method. In P. L. Munhall & C. O. Boyd (Eds.), *Nursing research: A qualitative perspective* (2nd ed., pp. 99–132). New York: National League for Nursing.

Giorgi, A. (1997). The theory, practice, and evaluation of the phenomenological methods as a qualitative research procedure. *Journal of Phenomenological Psychology, 28,* 235–281.

Moustakas, C. (1990). *Heuristic research: Design, methodology, and applications.* Newbury Park, CA: Sage.

Moustakas, C. (1994). *Phenomenological research methods.* Thousand Oaks, CA: Sage.

Ray, M. A. (1994). The richness of phenomenology: Philosophic, theoretic, and methodologic concerns. In J. M. Morse (Ed.), *Critical issues in qualitative research methods* (pp. 117–133). Thousand Oaks, CA: Sage.

van Manen, M. (1990). *Researching lived experience: Human science for an action sensitive pedagogy.* London, ON: Althouse.

van Manen, M. (2006). http://www.phenomenologyonline.com/inquiry/2.html. [downloaded April 3, 2006]

Reading Phenomenology

Clarke, M. (1992). Memories of breathing: Asthma as a way of becoming. In J. M. Morse (Ed.), *Qualitative health research* (pp. 123–140). Newbury Park, CA: Sage. (Original work published 1990)

Kelpin, V. (1992). Birthing pain. In J. M. Morse (Ed.), *Qualitative health research* (pp. 93–103). Newbury Park, CA: Sage. (Original work published 1984)

Smith, S. J. (1992). Operating on a child's heart: A pedagogical view of hospitalization. In J. M. Morse (Ed.), *Qualitative health research* (pp. 104–122). Newbury Park, CA: Sage. (Original work published 1989)

van Manen, M. (1991). *The tact of teaching: The meaning of pedagogical thoughtfulness.* London, ON: Althouse.

van Manen, M. (Ed.). (n.d.). *Textorium.* Retrieved May 8, 2001, from http://www.ualberta.ca/~vanmanen/textorium.html

Ethnography: Methodological Resources

Agar, M. H. (1986). *Speaking of ethnography.* Beverly Hills, CA: Sage.

Bernard, R. H. (1988). *Research methods in cultural anthropology.* Newbury Park, CA: Sage.

Boyle, J. S. (1994). Styles of ethnography. In J. M. Morse (Ed.), *Critical issues in qualitative research methods* (pp. 159–185). Thousand Oaks, CA: Sage.

Carspecken, P. F. (1996). *Critical ethnography in educational research.* New York: Routledge.

Ellis, C., & Bochner, A. P. (Eds.). (1996). *Composing ethnography: Alternative forms of qualitative writing.* Walnut Creek, CA: AltaMira.

Ellis, C., & Bochner, A. P. (2000). Autoethnography, personal narrative, reflexivity: Researcher as subject. In N. K. Denzin & Y. S. Lincoln (Eds.), *Handbook of qualitative research* (2nd ed., pp. 733–768). Thousand Oaks, CA: Sage.

Fetterman, D. M. (1989). *Ethnography: Step by step.* Newbury Park, CA: Sage.

Goldman-Segall, R. (1998). *Points of viewing children's thinking: A digital ethnographer's journey.* Mahwah, NJ: Lawrence Erlbaum. (See also website at http://www.pointsof viewing.com.)

Greenwood, D. J., & Levin, M. (1998). *Introduction to action research: Social research for social change.* Thousand Oaks, CA: Sage.

Hammersley, M., & Atkinson, P. (1983). *Ethnography: Principles in practice.* London: Tavistock.

LeCompte, M. D., & Preissle, J. (1993). *Ethnography and qualitative design in educational research.* San Diego, CA: Academic Press.

Reason, P., & Bradbury, H. (2001). *Handbook of action research.* London: Sage.

Schensul, J. J., & LeCompte, M. D. (Series Eds.). (1999). *Ethnographer's toolkit* (7 vols.). Walnut Creek, CA: AltaMira.

Spradley, J. P. (1979). *The ethnographic interview.* New York: Holt, Rinehart & Winston.

Spradley, J. P. (1980). *Participant observation.* New York: Holt, Rinehart & Winston.

Tedlock, B. (2000). Ethnography and ethnographic representation. In N. K. Denzin & Y. S. Lincoln (Eds.), *Handbook of qualitative research* (2nd ed., pp. 455–486). Thousand Oaks, CA: Sage.

Thomas, J. (1993). *Doing critical ethnography.* Newbury Park, CA: Sage.

Van Maanen, J. (Ed.). (1995). *Representation in ethnography.* Thousand Oaks, CA: Sage.

van Manen, M. http://www.phenomenologyonline.com/inquiry/2.html. [downloaded April 3, 2006]

Wolcott, H. F. (1999). *Ethnography: A way of seeing.* Walnut Creek, CA: AltaMira.

Reading Ethnography

Applegate, M., & Morse, J. M. (1994). Personal privacy and interaction patterns in a nursing home. *Journal of Aging Studies, 8,* 413–434.

Cassell, J. (1992). On control, certitude and the "paranoia" of surgeons. In J. M. Morse (Ed.), *Qualitative health research* (pp. 170–191). Newbury Park, CA: Sage. (Original work published 1987)

Davis, D. L. (1992). The meaning of menopause in a Newfoundland fishing village. In J. M. Morse (Ed.), *Qualitative health research* (pp. 145–169). Newbury Park, CA: Sage. (Original work published 1986)

Germain, C. (1979). *The cancer unit: An ethnography.* Wakefield, MA: Nursing Resources.

Gubrium, J. F. (1975). *Living and dying at Murray Manor.* New York: St. Martin's.

Morse, J. M. (1989a). Cultural responses to parturition: Childbirth in Fiji. *Medical Anthropology, 12*(1), 35–44.

Spradley, J. P. (1970). *You owe yourself a drunk: An ethnography of urban nomads.* Boston: Little, Brown.

Grounded Theory: Methodological Resources

Charmaz, K. (2000). Grounded theory: Objectivist and constructivist methods. In N. K. Denzin & Y. S. Lincoln (Eds.), *Handbook of qualitative research* (2nd ed., pp. 509–535). Thousand Oaks, CA: Sage.

Charmaz, K. (2006). *Constructing grounded theory: A practical guide through qualitative analysis.* Thousand Oaks, CA: Sage.

Chenitz, W. C., & Swanson, J. M. (1986). *From practice to grounded theory*. Reading, MA: Addison-Wesley.

Clarke, A. (2005). *Situational analysis: Grounded theory after the postmodern turn*. Thousand Oaks, CA: Sage.

Dey, I. (1999). *Grounding grounded theory: Guidelines for qualitative inquiry*. New York: Academic Press.

Glaser, B. G. (1978). *Theoretical sensitivity: Advances in the methodology of grounded theory*. Mill Valley, CA: Sociology Press.

Glaser, B. G. (1992). *Basics of grounded theory analysis: Emergence vs. forcing*. Mill Valley, CA: Sociology Press.

Glaser, B. G., & Strauss, A. L. (1967). *The discovery of grounded theory: Strategies for qualitative rsearch*. Chicago: Aldine.

Lee, R. M., & Fielding, N. G. (1996). Qualitative data analysis: Representations of a technology: A comment on Coffey, Holbrook and Atkinson. *Sociological Research Online, 1*(4). Retrieved May 8, 2001, from http://www.socresonline.org.uk/socresonline/1/4/ lf.html

May, K. (Ed.). (1996). Advances in grounded theory [Special issue]. *Qualitative Health Research, 6*(3).

Schreiber, R. S., & Stern, P. N. (Eds.). (2001). *Using grounded theory in nursing*. New York: Springer.

Stern, P. N. (1994). Eroding grounded theory. In J. M. Morse (Ed.), *Critical issues in qualitative research methods* (pp. 212–223). Thousand Oaks, CA: Sage.

Strauss, A. L. (1987). *Qualitative analysis for social scientists*. New York: Cambridge University Press.

Strauss, A. L., & Corbin, J. (1994). Grounded theory methodology: An overview. In N. K. Denzin & Y. S. Lincoln (Eds.), *Handbook of qualitative research* (pp. 273–285). Thousand Oaks, CA: Sage.

Strauss, A. L., & Corbin, J. (1998). *Basics of qualitative research: Techniques and procedures for developing grounded theory* (2nd ed.). Thousand Oaks, CA: Sage.

Reading Grounded Theory

Corbin, J., & Strauss, A. L. (1992). A nursing model for chronic illness management based on the trajectory framework. In P. Woog (Ed.), *The chronic illness trajectory framework: The Corbin and Strauss nursing model* (pp. 9–28). New York: Springer.

Glaser, B. G. (Ed.). (1993). *Examples of grounded theory: A reader*. Mill Valley, CA: Sociology Press.

Lorencz, B. J. (1992). Becoming ordinary: Leaving the psychiatric hospital. In J. M. Morse (Ed.), *Qualitative health research* (pp. 259–318). Newbury Park, CA: Sage. (Original work published 1988)

Morse, J. M., & Bottorff, J. L. (1992). The emotional experience of breastfeeding. In J. M. Morse (Ed.), *Qualitative health research* (pp. 319–332). Newbury Park, CA: Sage. (Original work published 1988)

Stern, P. N., & Kerry, J. (1996). Restructuring life after home loss by fire. *Image: Journal of Nursing Scholarship, 28*, 9–14.

Turner, B. A. (1994). Patterns of crisis behaviour: A qualitative inquiry. In A. Bryman & R. G. Burgess (Eds.), *Analyzing qualitative data* (pp. 195–216). London: Routledge.

Texts That Describe Several Methods

Brizuela, D., Stewart, J. P., Carrillo, R. G., & Garbey, J. (2000). *Acts of inquiry and qualitative research.* Cambridge, MA: Harvard Educational Review.

Crabtree, B. F., & Miller, W. L. (Eds.). (1999). *Doing qualitative research* (2nd ed.). Thousand Oaks, CA: Sage.

Creswell, J. W. (1998). *Qualitative inquiry and research design: Choosing among five traditions.* Thousand Oaks, CA: Sage.

Denzin, N. K., & Lincoln, Y. S. (Eds.). (1994). *Handbook of qualitative research.* Thousand Oaks, CA: Sage.

Denzin, N. K., & Lincoln, Y. S. (Eds.). (2000). *Handbook of qualitative research* (2nd ed.). Thousand Oaks, CA: Sage.

Ezzy, D., Liamputtong, P., & Hollis, D. B. (2005). *Qualitative research methods.* Oxford, UK: Oxford University Press.

Flick, U. (1998). *An introduction to qualitative research.* London: Sage.

LeCompte, M. D., Millroy, W. L., & Preissle, J. (Eds.). (1992). *The handbook of qualitative research in education.* San Diego, CA: Academic Press.

Marshall, C., & Rossman, G. B. (1999). *Designing qualitative research* (3rd ed.). Thousand Oaks, CA: Sage.

Maxwell, J. A. (1998). Designing a qualitative study. In L. Bickman & D. J. Rog (Eds.), *Handbook of applied social research methods* (pp. 69–100). Thousand Oaks, CA: Sage.

Morse, J. M., & Field, P. A. (1995). *Qualitative research methods for health professionals* (2nd ed.). Thousand Oaks, CA: Sage.

Rossman, G. B., & Rallis, S. F. (1998). *Learning in the field: An introduction to qualitative research.* Thousand Oaks, CA: Sage.

Rothe, J. P. (2000). *Undertaking qualitative research: Concepts and cases in injury, health and social life.* Edmonton: University of Alberta Press.

4

Qualitative Research Design

A common feature of qualitative projects is that they aim to create understanding from data as the analysis proceeds. This means that the research design of a qualitative study differs from that of a study that starts with an understanding to be tested, where often the hypothesis literally dictates the form, quantity, and scope of required data. This sort of design preempts other ways of looking at the research question.

Qualitative research is usually not preemptive. Whatever the study and whatever the method, the indications of form, quantity, and scope must be obtained from the question, from the chosen method, from the selected topic and goals, and also, in an ongoing process, from the data. Thus research design is both challenging and essential, yet it is the least discussed and least adequately critiqued component of many qualitative projects.

Freedom from a preemptive research design should never be seen as release from a requirement to have a research design. In Chapter 2, we established how a research purpose points to a research question and how the question informs the choice of method. But these choices do not remove the task of designing a qualitative project. Therefore we start this chapter by looking first at the levels of design and then at the goals of designing to specify the ultimate *scope* of a project and the *type* of data required. We end with practical advice on how you can tackle the ongoing tasks of designing your project so that you develop a research topic into a researchable question; we discuss the different levels and ways of planning, and the pacing of the project as a whole.

〽 THE LEVELⱭ OF DEⱭIGN

Research design is created by the researcher, is molded (rather than dictated) by the method, and is responsive to the context and the participants. Creating research design involves seeing the project at different levels. Once you have located your project methodologically, you need to design the *pacing* of processes and *strategies* to be used, and at the same time you need to see the project as a *whole*.

The *pacing* of the project involves planning the sequencing of its components and the movement between data gathering and data analysis. This requires ongoing decisions during the project: When should you stop interviewing? When do you return to observing—as processes of analysis show that more data are needed to verify, or when thin areas in analysis are revealed? The selection of method informs selection of research strategies, but these are also chosen in the context of the research question (i.e., what you want to find out) and the research context. For example, in studying the experience of menopause in a Newfoundland village, Davis (1983) relied on interviews rather than observational data. Richards, Seibold, and Davis (1997) were investigating the social construction of menopause, so they used observation of women's support groups and information centers as well as many forms of interview.

The overall design of the project must be aimed at answering your research question, and we look at detailed examples of design below. You need to design a project that both fits and is obtained from the question, the chosen method, the selected topic, and the research goals. You should treat research design as a problem to be considered carefully at the beginning of the study and reconsidered throughout—it is never a given.

〽 PLANNING DEⱭIGN

Where to start? If the questions, problem, and method are to guide design, then this becomes a highly conceptual and complex process. It is helpful to start with two questions: What is the *scope* of this project? and What is the *nature* of the data required?

The Scope of the Project

By *scope,* we refer to the domain of inquiry, the coverage and reach of the project. Scope involves both the substantive area of inquiry (the limits of the research topic) and the areas to be researched (the setting[s] and the sample).

Definitions of the topic and the relevant concepts and theories as perceived by other investigators in part delimit the area of inquiry. Consideration of the scope of the study continues in the process of gathering and analyzing data. You must work carefully and in depth, without losing sight of the research goals; remain flexible, self-critical, and, at all times, analytic; and use the literature as a comparative template. Coding decisions demand that you constantly ask: "Is this an instance of this category, or is it something different?" During the project, you must continually revisit the substantive scope of inquiry. If the data do not fit the question, analysis is likely to lack clear focus; the project may take too long to saturate and conceptualize and so, frustratingly, may achieve very little. On the other hand, if the scope is set too rigidly too early, the study will be severely limited. Avoid preemptively committing your study to definitions of the phenomenon of interest and concepts from the literature, thereby predetermining meanings of concepts; avoid making decisions too early in the study and drawing conclusions too quickly. Such preemptive scoping will result in premature closure.

The *scope of the sample* and the selection of the setting are driven by two principles. One is that setting and sample are purposively selected. This may involve choosing the "best," most optimal example of the phenomenon and the setting in which you are most likely to see whatever it is you are interested in. It may involve observing or interviewing experts in that particular topic or experience. Alternatively, you may select a setting because it allows you to obtain examples of each of several stances or experiences. The study may proceed by snowball sampling (seeking further participants by using the recommendations of those participants already in the study).

The second principle of sampling is that once you have begun to understand whatever it is you are studying, your sampling strategies normally are extended through *theoretical sampling* (Glaser, 1978). This means that your selection of participants is directed by the emerging analysis, and the theory being developed from data is subsequently modified by data obtained from the next participants. The scope of a study is never just a question of how many, but always includes who, where, and which settings will be studied; in what ways, by whom, and for how long they will be studied; and what can be asked and answered. All of these questions must be asked repeatedly as the project progresses. The research question may require that you seek out *negative cases* (examples of experiences that are contrary to cases that support the emerging theory, and that provide new dimensions, perhaps as indicated by the theory but not yet encountered) or *thin areas* from participants who have experienced special conditions that have been identified as significant. The scope of a project is bigger than its sample, for participants provide information about others *like them* or *unlike them*. Such "shadowed data" (Morse, 2001) provide you with further direction for your theoretical sampling. When sampling, you must be aware of when you are working inductively and discovering and when you are working deductively and verifying.

The interrelationship of the two components of scope becomes clear during the processes of data gathering and analysis. You need to ask constantly, "What scale of data and what range of settings and sources of data will give the strands required for *this* question, *this* topic, *this* method, *this* audience, *this* disciplinary or political context?" Asking and answering these questions about the project will help you to locate it, to establish the bounds of the question to be addressed and the goals to be rethought and realistically revised.

Designing the Scope

Scoping is an ongoing process in a project. It is rare for a qualitative researcher to set a scope and stick to it. Adjustments to the mode of making data are frequently required so that the project can be data driven. But this does not mean that such changes can "just happen." Changes ideally build upon the researcher's growing understanding of the situation.

We recommend that you always keep in mind the following issues regarding scoping:

⚒ The *substantive scope* of a project involves issues of comparison ("Will I understand the wider situation if I stay in this group?") and intervention ("Before I influence policy, how would I know if I were wrong?"). How many perspectives are needed? It is hard, for example, to study relationships only by observing interaction. If your question is about the relations between management and staff, you need to observe, if possible, but you must interpret your observations strictly in terms of your presence. You will also need other data sources; you need to talk to the managers and the staff, and you should examine relevant documents. These data sources will provide conflicting information—and you as the researcher have to make sense of the contradictions.

〰 *Scoping for change* involves asking if this is a study of a process (most qualitative studies are) and, if so, what time period it involves. Beware of studying a process with static data. One-off interviews, for example, will give interviewees' accounts, or the versions they see as appropriate in the interview situation, of what happened in the past. Is this the process and are these the perceptions you need?

〰 *Scoping for diversity* involves examining the sample, asking questions like "Is the research question comparative? If so, how do I achieve an adequate comparative base?" As you come to understand gender, race, or class divisions, new issues of scope will emerge ("If I observe only those folks, I will not be accepted over there"). Scoping for diversity involves considering the scale of the research question ("Whose experience will I not hear?"). It requires attention to representation ("What is it that I want to make statements about? Does what I won't see matter?"). It also requires attention to the areas to be covered ("Is there more than one perspective on this issue?").

As you reevaluate each of these issues, the answers will shift in response to your discovering, theorizing, and constructing theory. Scoping the project almost always shifts the question in the interplay between what can realistically be asked and what can properly be discovered. The process moves the question from a research question to a *researchable* one.

The Nature of the Data

How will you create data, and how will you ensure a fit of data to the research task? These are different questions. They require you to explore the possible ways of constructing data within a setting and to select methods that will combine to ensure that the data will be sufficiently rich, complex, and contextual to address the question and support the required analysis.

Thus, rather than preparing a research instrument for use throughout the project, in undertaking the design of a qualitative study, you need to consider carefully the variety of approaches available and the sorts of data they generate. Predesigned research instruments may be useful for some tasks (e.g., a survey form may be used to record basic demographic data about participants). But because the goals of the project include learning inductively from the data, instruments designed entirely in advance will rarely support an entire project.

You should expect that an interesting research question will usually require several strategies for making data. Relying on one technique may produce homogeneous data, which are highly unlikely to provide enough sources of understanding and ways of looking at a situation or a problem. Commitment to one sort of data makes the techniques of theoretical sampling very difficult to follow, so you need to resist the easy route of selecting one technique and building in the assumption that you will "do focus groups" or "do in-depth interviews." Keep asking, "Why would this one way of making data suffice to answer my question?" We share a concern with other scholars regarding the increasing homogeneity of data in qualitative projects as the dominant mode of uniform, as "in-depth" interviews take over from the previous speckled diversity of qualitative data. Our advice is that you not look first for a technique of making data with which you are familiar or that you have been trained to do, but rather ask how, in this situation, you can best access accounts of behavior and experience, best weigh the different versions of "reality," and best interpret them.

You should expect that the nature of the data will change during your project. The importance of knowing a budget and timeline can easily overtake the requirement of growing a project informed by the data. Starting with the assumption that they are "doing interviews," researchers are easily led to see as the only relevant question the issue of how many respondents they should "do." (We recommend reflection on discourse here—both about what you are proposing and how you are expressing it.) Even the most expert researchers cannot answer the sample size question without involvement in the project. What constitutes a large enough sample will be determined in the future by the situation studied and the quality of data. But the fact that the question is asked should alert you to its corollary: "What else could or should I be doing to create a strong and rich data set?"

Focus on the end, not only on the beginning, of the project, and particularly on the claims to be made ("What am I asking of these data?" "What types and combinations of data do I need to create?"). Try to foresee the adequacy of likely results ("What will I not see if I rely on these sources?").

Ask yourself about your own ability to create the data ("Will I be able to do this, to be accepted in this situation, to conduct these sessions, to find participants?"). Try also to foresee limitations ("If I seek nuances of meaning in people's language, how do I ensure that my records contain these and that they are not determined by my intervention?"). At this earliest stage, it is helpful to *think backward* from possible outcomes. What sort of a study of this issue would be convincing? What ground do you want to be able to claim? Who do you want to persuade and how would they be persuaded? How will you know, at that wonderful final stage of reporting, if you were wrong?

〽 DOING DEJIGN

We have emphasized the importance of allowing the questions, problem, and method to inform the scope of the project and nature of the data, and also the importance of the researcher's actively designing and controlling the project. How do you do both?

A good place to start is to read other studies critically. What is it about particular studies and their designs that convinces you (or that is convincing to you)? Do those authors persuade you that they were not wrong? The qualitative studies that you find exciting are likely to be convincing because the projects had the scope of design and the nature of data necessary to answer the research questions with the methods chosen. Unconvincing projects are those in which the researchers try to make claims where there is no justification or try to stretch thin data beyond their capacity to hold an argument.

If the task of starting is daunting, we recommend that you approach it by taking the five steps outlined below. As you prepare your proposal, you will find it helpful to keep an account of these steps and your thinking as you proceed, and of the puzzles that confront you and the ideas that occur. Many researchers commence their projects with proposals that avoid critical questions, which also often means that they avoid design—a very problematic stance.

〽 *Step 1: Establishing purpose.* What are you asking? Why are you asking it? Who has asked it or something like it before, and how and why did those studies not satisfy your curiosity? (Treat your literature review as qualitative research.) What are you doing that adds to what they did? What is your intent? What do you want to come out of this? What do you know, and what advantage and disadvantage is this? Revisit the discussion of topic selection in Chapter 2. (Particularly, at this stage, do not assume that being "one of them" gives you enough knowledge to research "them." Treat being one of them as a problem, not an advantage.)

⋙ *Step 2: Methodological location.* What is the appropriate fit of qualitative method to this question and topic? Never start with the method and then seek a topic. Does the method point you in the direction of research design? Particular methods usually require certain sorts of data—what sorts of data are you going to need to do your project this way? Revisit the discussion of the armchair walkthrough technique in Chapter 2.

⋙ *Step 3: Scoping.* Now move on to the task of defining the scope of your project. What is it that you want to make statements about? Do you know enough about the field to determine who you should sample? If not, build in preparatory fieldwork—this is *not a pilot* but a stage in itself. Do you know enough about the issues? If not, build in preparatory identification of them. Are you comparing anything? If so, design for comparison. Are you intervening? If so, how, and are you planning for this?

⋙ *Step 4: Planning the nature of your data.* What *sorts* of data will be relevant? What sorts are available? How, and in what order, will they be combined? Are you able to handle those sorts of data? The design should include your data-handling methods and the ways you will use software. (Note that one of the classic howlers of research is to say of any software program that it "will analyze" the data!)

⋙ *Step 5: Thinking ahead.* How satisfying will this study be? How robust? Why should it be believed? How will you know if you were wrong? Present your proposal to skeptical audiences and become a skeptic yourself. The goal is to start your study knowing that it will be convincing at the end.

⋙ DEƒIGNING FOR VALIDITY

Validity is a term too often avoided in qualitative research, because it is mistakenly seen as an indicator of attitudes to analysis or to interpretation that do not fit with qualitative methods. In the literature of every method you will find debate about the term's possible meanings in qualitative research, and sometimes alerts about "the crisis of validity" (Denzin & Lincoln, 2000) or complex suggestions about specially "qualitative" terminology. As you prepare your research design, it is important to be aware of how the issues are considered in your chosen method.

However, it is also important to ensure that you are designing a project whose outcome will be appropriate and fully justifiable, as properly based in the data. This is the commonsense and dictionary meaning of "validity": a valid assertion is "well founded and applicable; sound and to the point; against which no objection can fairly be brought" (Shorter Oxford English Dictionary, in Richards, 2005, p. 139).

Two general rules guide research design for validity in all qualitative projects. The first is the theme of this book: Pay attention always to the fit of question, data, and method. This will ensure that the data are appropriate and appropriately handled and the question addressed fully and responsibly. From this requirement, it may follow that you should set up specific ways of checking how the data and method are performing. For example, it may assist a team project to check the reliability of coding. However, those checks should always be designed and carried out consistently with the method (Richards, 2005, pp. 139–144).

The second general rule is to ensure that you can properly account for each step in your analysis. All qualitative projects get their claim to being trustworthy from the ability of the researcher to account for the outcome (Maxwell, 1992). From this requirement, it follows that from the design stage, you should set up processes by which you can log each significant decision and the interpretation of each discovery. To do this as you work will be very important. Remember that qualitative analysis builds theory out of the data, one interpretation providing the platform for another inquiry. Your log of that journey will be the prime source of your justification of where you arrived and what you discovered (Richards, 2005, pp. 22–26).

At the research design stage, consider what your project design needs now to ensure that your conclusions will be regarded as sound and well founded. The steps outlined in this chapter take you through stages of design, with warnings about how they can go wrong. For more detail on the ways you can check the soundness of your analysis, go to Chapter 9 where we return to the challenges of "getting it right."

❧ PROJECT PACING

What does a good design look like? The sort of evolving design described above will be less tidy than one for a survey research project, where properly collecting data is the first stage, followed in turn by coding and analyzing. A qualitative research design is more like a journey in which each of the stages builds on previous experiences. Planning flexibly for these stages helps you to confront the work that is often not factored into a design, to budget time and money, and to distribute workloads and manage relationships with your significant others. One of the interesting results of planning this way is that you discover that no stage ends neatly so another can begin. Whether or not you are required to make a formal proposal with timelines, it is worthwhile to draw up a schedule that includes the five stages described below. (If you are required to write a proposal, see the details provided in Chapter 11.)

Conceptualizing Stage

Plan and budget for careful thinking through of the project, the literature review, and critiques of other studies. Plan to do this early—and keep doing it. You will need to continue to read and critique the literature throughout the study as new relevancies appear and new studies emerge. Handle the literature review data as data—using the data-handling method you intend to use later for the interviews or field notes. (If you are using a computer program, now is the time to get to know it well. Tutorial 1 on the CD shows the basic techniques for storing the materials acquired at this stage.)

Entering the Field

Treat entering the field as research work: Prepare and budget for it. Your field may be a location (such as a school), or it may be entering a topic (the people who have the disease you are studying or who share an experience of discrimination). In many disciplines, the emphasis is on making data through direct, obtrusive methods such as interviews or focus groups, where researchers are deprived of the insights of ethnography. If you do not know the literature on field research, explore it now. It will alert you to the observer's task of preparing for, gaining entry to, and becoming accepted in a setting.

If you are working in a familiar area, be especially careful. A useful mind-set is to regard yourself as reentering the field as an observer. Assume that the advantage of understanding problems and perspectives is at least partly balanced by the disadvantage of the insider's taken-for-granted assumptions, commitments, labels, and ways of seeing. If you are studying a familiar topic (a problem or group you know) by more obtrusive methods, such as interviews, be particularly cautious on entering the field. When you spend only a couple of hours with an interviewee, your assumptions can go unchallenged.

Setting Up and Managing a Data Management System

For obvious reasons, any research design must include the ways in which you intend to handle data. We hope that by now it is clear that selection of a data-handling system must be done very carefully: The system you choose must be tailored to the task and adequate to the scope of the project and the varieties of data and analysis expected. (The literature is often silent on this essential stage, but for exceptions, see Dey, 1995; Lofland & Lofland, 1995.) You need to plan for the data-handling system you will use from the beginning of the project; you must be sure that you are familiar with it and that it is working from the start. Note that this advice is especially important for team projects (Richards, 1999b, 2005). Your research design should allow for the time you will need to develop a system that works for you and for the time it will take for you to learn any skills, particularly computer skills, that the project will require.

Now is the time to learn the software skills that you will need throughout the project. You should at this stage make decisions about the software you will use, learn to use it competently, and become familiar with the range of processes it will support. Then, as soon as possible, you should start using it. To delay working with your software risks serious disadvantage. If material piles up on paper, waiting to be entered on your computer, the workload of managing your data will grow, as will your anxiety about being able to handle your data. To bring the early material immediately into the computer will give you confidence, time to learn software techniques, and the ability to integrate research design materials with the data records you will soon start to create. This chapter, and each of the next three, concludes with a section on software tools relevant to the chapter's content.

Sampling and Theoretical Sampling

Allow time in your design for the process of locating and evaluating the ways you can sample the studied area. This can be very demanding; never assume a sample is waiting for you like an apple to be plucked from a tree.

Treat theoretical sampling (i.e., the selection of participants according to the needs of your emerging analysis) as a necessity and build time and budget for it into your design. In a grant application, state areas where further sampling is likely, and budget time and other resources accordingly.

Analysis

Any project design must allow for the cognitive processes of research. Build thinking into your timelines and your budget. In a grant application, allow time for coding of data, for recoding of exploratory categories, and for management and exploration of category systems as well as for coding validation and reliability exploration. Allow time for asking questions and incorporating the answers into your analysis. And, above all, allow time for writing, rewriting, revisiting the data, and verifying your conclusions.

⦚ CHOOJING YOUR JOFTWARE

Researchers who have not previously used specialist qualitative software will face an obvious early task, to choose from the available software types and products. If the choice is not made for you, by institutional licensing or availability of skilled assistance, you will need to research the range of software functions, and then their appropriateness to your research design.

If you are planning to combine qualitative and quantitative modes of analysis, this means you are choosing *two* software packages. To make that choice requires a good knowledge not only of how each works but also of their compatibility and appropriateness for your project. Note from Tables 4.2 and 4.3 that qualitative programs differ greatly in the ways they can link with quantitative programs.

Think of this choice process as another step in the pursuit of methodological congruence. Just as research purposes and questions fit with data types and analysis strategies, so do software tools fit, for better or worse, with all these aspects of research design. Start there, setting out what you are asking, what data you expect to be handling, and by what methods of analysis, and from there ask which of the tools available in software would best assist you.

The task is less daunting now than it was in earlier stages of software development. When computer tools were first designed for qualitative research, very different types could be identified (Tesch, 1990; Weitzman & Miles, 1995). Two decades later, there is a substantial common ground for basic functions, summarized in the regularly updated comparisons at http://caqdas.soc.surrey.ac.uk/, the website of the CAQDAS Networking Project (the acronym is for Computer Assisted Qualitative Data Analysis).

The main commonalities and differences at the time of writing are summarized in the following tables.

Table 4.1   Your Project and Your First Data

Expect This of All Qualitative Software	Look for These Differences	When Will This Matter?
Provision for storage and managing of data and interpretations in a single unit or project	Programs differ in how a project is saved, stored, and transported.	Researchers nervous about security, ease of backup, and sharing of team projects should check for software help.
Ways of combining and comparing projects	Most programs support merging of projects, but they differ in flexibility of merging parts of projects.	This will matter if you have multiple researchers or sites or a good reason for combining your own projects.
Ways of backing up and safely storing projects	Programs differ in whether the source data is imported in the software or remains external to the software.	Mode of storage if not understood can imperil a project—be very clear about what should be backed up!
Ways of interfacing with other software	Programs differ greatly in whether they are designed to import from and export to statistical or database software.	This will be critical if you plan mixed methods research.

Table 4.2   Your Data Documents, Ideas, and Links

Expect This of All Qualitative Software	Look for These Differences	When Will This Matter?
Handling of text data for the project prepared in word processors	Programs differ in permitted format of files (plain text, rich text, or word processor formats) (including pictures, tables, etc.).	Text formatting matters most for projects with "rich" records.
Ability to create and edit text within the project. All will allow typing of memos.	Programs differ in whether data documents can be freely edited once "in" and in the flexibility of editing memos.	Typing up in the project matters most for records you want to code or annotate as you create them.
Inclusion of text data files in the project	Programs differ in whether they import the documents or link to files kept externally.	Where the data are stored may matter for security and convenience.
Handling nontext data—photos, videos, etc. (either importing and coding directly or ways of representing nontext records)	Some but not all will import nontext (pictures, video, or audio). Others vary in terms of what can be done with multimedia data.	A central concern if your design requires detailed analysis of nontext data.
Storing information (such as demographics) about people or places, etc.	Most will import such information from spreadsheets or statistics software—but differ in options and display—and the flexibility with which you can use this information.	This is critical if you are doing mixed methods research or have a large sample.
Creation and editing of documents and memos from within the program	Programs differ in flexibility to edit and in ways memos are created and whether and how they are searched. Check whether the program is designed for extensive editing or just for corrections.	This is important if your method requires constant reflective records as theories are built (e.g., grounded theory).

Table 4.2 (Continued)

Expect This of All Qualitative Software	Look for These Differences	When Will This Matter?
Annotating or commenting text	Programs differ in ways of annotating particular passages of your data, how annotations can be viewed, and how they are reported.	This is important if your method requires fine detail commenting of discourse in texts (e.g., discourse analysis).
Support for linking to data within project and outside	Very different approaches and methods of linking: Look for what can be linked and how.	This matters if your method requires you to bring data together in ways other than coding.

Table 4.3   Coding and Text Search

Expect This of All Qualitative Software	Look for These Differences	When Will This Matter?
Coding of selected data at categories created by the researcher (called "codes" or "nodes") and retrieving all data coded at a category	Programs differ in mode of selection of data and procedure of coding. Some programs allow the researcher to record weighting of coding.	Coding style and facility are important to most researchers—try out software to see if you like the way it codes!
Ability to view all of the data you have coded at a category	Programs differ in how you view coded data and whether and how the context can be retrieved. They also differ in how you can work with coded data to revise coding and optionally code further from it.	This matters most for methods where coding is just a first step toward interpretation, especially if it is important to explore the dimensions of a category.

(Continued)

Table 4.3 (Continued)

Expect This of All Qualitative Software	Look for These Differences	When Will This Matter?
Ability to see on the screen what coding you've done (usually in margin, sometimes by highlighting or reporting)	Programs differ in whether all coding can be seen at once and how the markings can be used to explore that code.	It will matter if you rely on (and are concerned about) coding or if in teams you want to compare coding.
Auto-coding of data (mechanical finding and coding of words or segments)	Programs differ in how easily this is done and how much formatting is required, as well as whether you can set the context you want coded.	This is particularly important to projects with a lot of very structured data or requiring immediate retrieval (e.g., everything said by a particular speaker).
Text search of words in data and sometimes coding of the finds	Programs differ in the ways they conduct searches and store results, and whether you can save searches and results.	This is important if your method requires the mechanical processes of word search and/or further exploration of results.
Counting of codes or occurrences of words; quantitative content analysis	Some offer word frequency counts and quantitative reporting of searches. Ability to create your own dictionary in some.	This is important if your method requires counts. If so, check out "text retriever" and "content analysis" programs.

Table 4.4 Abstracting, Modeling, Questioning, and Reporting

Expect This of All Qualitative Software	*Look for These Differences*	*When Will This Matter?*
Management and viewing of coding categories	Most programs support hierarchical cataloging of categories for review and access.	This is important if your coding categories will be numerous and/or if you are sharing coding schema.
Asking questions (with "search" or "query" tools) about patterns in the coding of data	Programs differ in ability to search combinations (e.g., Boolean, proximity) of coding, text, and the characteristics associated with people or places. Some allow multiple searches producing matrices for pattern exploration.	Some methods require sophisticated searches (e.g., matrices to show patterns). Do a "walkthrough" of the project to check how you want to query your data.
Saving of search results	Some save them as reports only, while others allow the possibility for the search results to be incorporated in the database.	This matters if you want to build inquiries on your coding patterns.
Ability to run repetitive searches	Only some programs provide for the researcher to write scripts to set up analysis processes.	This is important for projects where computer searching must be adapted to the design.
Visual displays	Most provide some tool for modeling. These vary greatly from simple diagrams to live-to-data representations of theories and networks.	If visual representation of what is happening in your project is important, check what you need the software to display.

(Continued)

Table 4.4 (Continued)

Expect This of All Qualitative Software	Look for These Differences	When Will This Matter?
Ways of seeing connections you have recorded in the data	Some packages have tools to get you "up" from the text and display connections in a model view.	This is important for analyzing in depth a case instead of across cases and for getting the Big Picture of a project.
Making reports of data, codes, coding, etc.	Programs vary in the way reports are created and presented.	If you have particular report needs, check that these are supported.

╲╲ OVERALL PROJECT DESIGN

We return, in conclusion, to the top level of design—your vision of the overall project. Novice qualitative researchers are sometimes prevented from taking an overall view of their projects because they are so insistently told not to be preemptive. There is danger here that you will never see what you are aiming for, and without such a vision, design is impossible. Flexibility and willingness to learn from the data do not require lack of purpose. The ability to see and design the overall project is particularly important in comparative research and in projects that combine different sorts of data. We discuss two such design challenges below: comparative design and triangulated design.

Comparative Design

If your question demands that you determine what is special about a group or identify particular conditions or circumstances, then you may need a two-group design. Ideally, you will keep data from the two groups separate, as well as theoretical sampling driven by each group, and the data will be saturated separately. Later in the research process, you will compare and contrast these data sets to determine similarities and differences between the two groups. Sampling and data collection will continue beyond this phase so that the emerging analysis may be expanded and verified.

Comparative qualitative research is important, for example, in evaluation. Qualitative inquiry alone will not answer "how much" or "how many" or, therefore, how great a difference has resulted from an intervention. But a researcher can look at such questions by using a two-group design in which both groups have had similar experiences but only one group received the intervention, or by examining a single group before, during, and after intervention. Such studies, conducted outside of the laboratory, are called naturalistic experiments.

Triangulated Design

A text search of grant applications in many countries would return a high count for the odd word *triangulation*—so would a vote among researchers for the most misused term. Originally coined to describe a specific sort of research design, *triangulation* is now widely used to mean vaguely "three sorts of something." But exactly what is triangulation?

Triangulation refers to the gaining of multiple perspectives through completed studies that have been conducted on the same topic and that directly address each other's findings. To be considered triangulated, studies must "meet"—that is, one must encounter another in order to challenge it (for clarification), illuminate it (add to it conceptually or theoretically), or verify it (provide the same conclusions). Goffman (1989) coined the term, drawing on the metaphor of the surveyor's practice of making sightings from two known points to a third.

Like the surveyor, the qualitative researcher may be aided by drawing from different perspectives on the same question or topic. Triangulation requires careful research designs to ensure the same question will be addressed, and answered, by each of the proposed approaches (Richards, 2005, p. 140). A researcher may do this by juxtaposing analysis of different data *types* and *methods* to illuminate the same *question* (e.g., field note records of participant observation in a school are examined for the picture they give of authority behavior, and authority is "sighted" differently via a study that used interviews in private with students and teachers). More ambitiously, researchers may address the same question in *separate studies* designed for direct comparison and using different methods and data. If they reach the same conclusion, they will use triangulation to verify or challenge alternative interpretations. Or, in a third form of triangulation, a researcher may address the same topic as that addressed by another, but through a different *question, method, setting,* and *data* to gain a different perspective (e.g., in a study of prejudice in a community, a study using a survey method gathers precoded results to feed statistical analysis, while observation in various community organizations reaps a record of how people from the same community, who may or may not have participated in the survey study, behave in public).

All of these are useful techniques—as long as the question is being properly sighted from the different angles. Such multiple sightings on an understanding can be extraordinarily revealing; a new interpretation of the sources of teachers' apparent authority may come to light, or new information concerning the ways in which prejudice is concealed. The necessary condition of such good results is, of course, a good research design. If you elect to use a triangulated design, it is essential that, from the start, you are clear about your purpose in using such a design, and that you make explicit in each stage of the research design what you are claiming you can do with it and what it will add.

Avoid misusing the term *triangulation*. One does *not* do triangulation by interpreting the same data using different theories, or by gathering a multidisciplinary team of investigators or coders. And although you may have complicated data—for example, in the form of a couple of data sets, neither of which can be a complete study in itself, or a few interviews conducted from a different perspective—they do not necessarily offer different sightings on the same question. The following research designs, all of which may be constructive and successful, are *not* examples of triangulation:

⁂ Multiple data sources used in a single study to build a single picture (such as ethnography consisting of interviews, observations, and so forth)

⁂ A second study added on to the first (six case studies more!) (Increasing your sample size is not triangulation.)

⁂ Two studies that do not reflect on the same phenomenon or the same question (Studies that are not designed to give sightings on the same question are highly unlikely to meet in their conclusions.)

⁂ Juxtaposition of quantitative and qualitative inquiry (For example, a study of stepparenting that uses in-depth interviews of stepparents may be informed by a study using census data on the distribution of stepparenting across California. The survey provides context for the qualitative study, but it does not address the same question, nor does it strengthen, add to, or potentially challenge the results of the other.)

≫ COMBINING QUALITATIVE AND QUANTITATIVE METHODS

When the research problem is complex or if the researcher suspects that one method or strategy may not comprehensively address the research problem, multiple research methods may be used (i.e., multiple-method design) or a second research strategy may be used to supplement the core method (i.e., mixed method design). The combination of methods is less likely to threaten validity in multiple-method design as each method is complete in itself.

Mixed Method Design

In mixed method designs, a single method and one or more strategies drawn from a second method are used in the same project—these may be both qualitative, one qualitative and one quantitative, or two quantitative (however, the last design, two quantitative, are beyond the scope of this book). Mixed method designs are usually used because one method alone will not provide a comprehensive answer to the research question. Perhaps in a study that is primarily quantitative, there is some aspect of the phenomenon that cannot be measured; or in a study that is primarily qualitative, there is some aspect of the study that can be measured quantitatively, and the measurement will enhance our descriptive understanding of the phenomenon. Or perhaps if two qualitative methods are used, one will complement the other, for instance, provide access to a perspective that cannot be accessed by the first.

Morse, Wolfe, and Niehaus (in press) argue that several principles are important to attend to when conducting a mixed method project.

≫ Analytic weighting of both components: Both components are not equally weighted in the analysis; one component provides the analytic core, and the other component's findings add to or fit into that core or provide insight adding explanation to the findings of the core component.

≫ Completeness of components: The core component is complete in itself, and the supplemental component is not complete enough to stand alone—that is, it is not publishable as a separate component. The supplemental strategies may be, for example, data obtained from several focus groups from which items are developed for a quantitative questionnaire, or quantitative measurement of anxiety to describe the *anxiety levels* of participants within an ethnography of relatives' waiting rooms in a hospital.

⧼ Theoretical drive: Mixed method projects are either inductive or deductive, as determined by the research question, and subsequently by the primary or core method. The overall inductive or deductive direction of inquiry is referred to as the *theoretical drive* of the project, and the major methodological assumptions must be consistent with the assumptions of the major method. For instance, if the core project is quantitative, then the core design and the analytic methods are quantitative, and vice versa for qualitative. The theoretical drive of the project is indicated in the sequencing below in caps, for instance, QUAL for a qualitative theoretically driven project and QUAN for a quantitative theoretically driven project. The notation for the supplemental component is lowercase letters.

⧼ Sequencing: The sequencing of the project is important. The core and the supplemental project may be conducted at the same time (*simultaneous*, indicated with a +), or with the supplemental project following the core project (*sequential*, or indicated with an —>). Simultaneous mixed method design may permit the transformation of the qualitative data to quantitative numerical data and incorporation into the quantitative data set. But this will be possible only if the qualitative methods used ask the same question of all participants and the sample size is equivalent to the quantitative sample, as, for instance, in the case of semistructured interviews, questions contained within a survey.

This means that we have the following combinations of mixed method designs:

QUAL + quan, a qualitatively driven project, with a qualitative core and quantitative simultaneous component

QUAL —>quan, a qualitatively driven project, with a qualitative core and quantitative sequential component

QUAN + qual, a quantitatively driven project, with a quantitative core and qualitative simultaneous component

QUAN —> qual, a quantitatively driven project, with a quantitative core and qualitative sequential component. An example may be conducting a quantitative survey and obtaining surprising results. A qualitative study is then conducted to obtain some insight into those results.

QUAL + qual, or QUAL —> qual, a qualitatively driven project with a qualitative core and a qualitative supplemental component, first conducted simultaneously and second, sequentially. An example of these designs might be a grounded theory project explaining the process of some aspect of parenting, and a phenomenological component used as a supplement to illustrate the meaning or essence of some part of the process, either simultaneously or sequentially.

〰 *Sampling:* Difficulties occur if the supplementary component is qualitative and the quantitative core sample is too large to incorporate the qualitative component. How, then, does the researcher select the qualitative sample? When this problem occurs, the qualitative sample is purposefully selected from the quantitative sample, or a separate qualitative sample is drawn, consistent with the principles of qualitative sampling. Conversely, if the core component is qualitative and the supplementary component quantitative, then the qualitative sample is too small for quantitative analysis. If it is necessary to measure some aspect of the qualitative sample, then the quantitative measure must have external norms against which to compare and interpret the scores obtained. Alternatively, a large quantitative sample must be drawn. Occasionally a mixed method project is designed with both a quantitative questionnaire and qualitative, opened-ended questions as a part of the questionnaire. Then, the qualitative questions are coded and transformed numerically and incorporated as variables into the quantitative data set.

Multiple-Method Design

In multiple-method research, the two (or more) projects may both be quantitative, both qualitative, or one qualitative and one quantitative. In this text, those methods that are both quantitative are beyond our scope, and we will discuss only those projects that are qualitatively driven and conducted concurrently or simultaneously within either a quantitative or a qualitative research program. Because both projects are complete in themselves, these projects are usually a part of programmatic research. That is, they are a part of a series of funded projects around a common research problem, and may only be located in *review* articles that examine the work of a single investigator or team.

The overall theoretical drive to the program may be identified by the programmatic overall question. Projects may be qualitatively or quantitatively driven, and the individual projects within the program may be qualitative (to provide meaning and insight) or quantitative (usually to test emerging ideas). However, it is important that the researcher recognize the *theoretical thrust* of the overall program, for that will enable the identification of the study that will form the theoretical base of the project. It is into that base that the *findings* of the other methods fit and inform the overall emerging model.

Note that because the projects are independent, it is the *results* that are combined in multiple-method research, not the data, nor the analysis. Therefore, multiple-method research escapes some of the quandaries of mixed method research with regards to the provision of appropriate and adequate samples and the transformation of data for analysis.

How are the results combined to inform the base project? Simply put, the results are combined in the process of writing in the form and format that will provide the reader with understanding. Because the numbers of studies are often more than two, the researchers often write monographs, with the previous studies reprinted from refereed journals, so the reader can assess the contribution of each study.

Taking an Overview

Seeing your project overall will prevent you from going into a situation with the conviction that understanding will just happen, or from collecting the data and then thinking about them. It will help you to avoid narrow designs, homogeneous data projects, and making data (often volumes of data) without being sure you have the skills and resources to handle the resulting richness. It will warn you against writing a proposal that gives a vague direction—to "get in there and find out what's going on." It will push you to sharpen your reflection on the appropriate *ways* to get in, on where or what *there* is, and how one *would* find out. The vaguer the research question and the less located the context, the more the project is at risk of wandering aimlessly—and the more you need a research design. Good studies rarely, if ever, just happen (and studies are not often funded on the basis of a promise that they will).

When detail threatens to cloud your bigger picture, return to looking at your project overall. Unfunded research requires focus on a question and a design to answer it, perhaps more urgently than does funded research, because the constraints may be greater. Just as the funding body needs to understand your research design, so do you need to know that your project is likely to contribute to understanding of your topic, that it is within your capability and resources, and that it has a shape and a likely outcome. Like the funding body, you will also want to know for ethical reasons that the possibly invasive processes of data making are designed to contribute an answer to something worth asking. During this process, you will be asking yourself constantly how to scope the project to maximize the chances of achieving an adequate answer, and how to design the data so that they contribute to an understanding that is not just good enough but convincing.

⟨⟨⟨ UJING YOUR JOFTWARE FOR REJEARCH DEJIGN

We end this chapter, and each of the next three, with a brief discussion of what specialized qualitative software offers for the tasks discussed, and how it changes what the researcher can do. These sections describe and discuss the ways of doing qualitative research using computer software under three headings—approaches, advances, and alerts. They advise on what you can ask of your software, then on the ways software enables doing things that could not be done when only manual methods were available, and finally on avoiding temptations and pitfalls.

All qualitative researchers now use software of some sort. Their interviews are typed on word processors, their sample characteristics in spreadsheets. But most will also use specialist software designed to help with the challenges of managing complex qualitative data records without losing their context, and storing and exploring the growing ideas about those data records.

What does software offer to the researcher approaching the issues discussed in this chapter? At this early stage of a project, what approaches are supported by software? What advances in method does software offer? And what are the traps and temptations to which you should be alerted?

Approaches

Early in a project, important tasks are organizational. Taking the steps to a good research design as described in this chapter, you will make a lot of records of your thinking, reading, scoping of the project, and planning. Research proposals, grant applications, and literature reviews are the intended outputs. The inputs can be messy and confusing.

Qualitative software is designed to help you handle messy inputs, so it will assist with these early records. Novice researchers often are misled into thinking they have to have "real" data before they can start using software. But there is no need to start by storing research design records separately from later records of interviews or field research. If you do so, it will be much harder to access these together throughout the project.

Start by learning how your software will store records and allow you to see them separately—in folders, sets, or groups. If you set up a project carefully, you will be able to access your plans and reviews alongside the other data records you will create when you commence interviewing or field research. So the contribution of software at the research design stage is as a reliable (but not rigid) container for plans, early considerations, and topics.

You need to learn now

〽 What your software can do, and how to do it

〽 How to start a project

〽 How to manage it—saving, backing up, and transporting it

If you wish at this stage to try working in qualitative software, go to Appendix 1, where there is a guide to tutorials available online.

If you are planning to combine qualitative and quantitative modes of analysis, it is important to build into your research design consideration of the ways you will "mix" these. To do so will involve moving data between software packages. While it is still common for a "mixed method" study to be designed simply as two studies (Bryman, 2006), good mixed method research does not merely juxtapose two projects but integrates them. To do this, you will need to plan, from the start of your research design, for the appropriate data and staging of analysis. In her full review of recent literature and methods, Bazeley (in press) distinguishes between two major strategies for integration: using software to combine numeric and text data and using software to *convert* coding from qualitative data for statistical analysis. This paper and other resources for mixed method research are available at www.researchsupport.com.au.

Advances

Computer software cannot design your project, but it can assist greatly in the data management tasks at this stage.

Once you learn software skills, you will find that starting early in software has great advantages. And more important, starting early in software does not disadvantage you. Software tools are now far more fluid than those first developed for qualitative research. A good software package will allow you to create a project and then later change practically all aspects of it, as your ideas about the data and analysis grow.

At the early stage of design, you can store drafts and estimates of project stages, and using the tools taught in later tutorials, you will be able to link them and shape the ideas that inform your design. As you work the ideas and issues, you will be able to see more clearly what design decisions must be made or how, for example, you can design the sample of your study to encounter the range of discovered issues.

In the early stages of research design, the computer offers storage for documents and for ideas—and the ability to link them by coding the relevant passages of documents at the relevant ideas, so that all the relevant material can be retrieved later. The research design can be informed and directed by systematic storage of early explorations of the topic, serious reflection on the range of options for approaching it, and informed decision making.

Alerts

⧂   Software is not a method. Having chosen the software you will use, ensure that all your research moves are directed by your design and your method. Always be concerned if you are doing something because your software can do it!

⧂   Starting in software gives great advantages later—so long as you ensure that you remain flexible. Where you start will not be where you go: This is built into the method. If you start your project with software, start flexibly. Use software tools for storing your *changing* definitions of concepts, finely coding data about them, ordering them, and exploring their relationships. Any software package will allow the coding categories to be changed at any time, reordered at any time, combined, or deleted as the data direct your understanding of them.

⧂   Keep moving! Getting out ideas, thinking about them, and making them accessible do not require a computer, but, like so many other tasks, it is dramatically easier and faster with the proper software. The risk is that this may encourage you to work on those early tentative ideas so long and so well that you start constructing the framework for analysis preemptively.

ℵ   Setting up a project is not analysis. You must also start and continue reflecting on what you do. Making and designing a project should involve processes of recording and logging your thinking about your research design. To do this on the computer is not necessary, of course, but it will help you to clarify the choices you have and the decisions to be made. As you prepare designs and time estimates, edit them to reflect, change, and manage them. As you store your early ideas, describe and write about them.

ℵ   Beware of flexibility as well as of rigidity! Don't allow the fun of setting up a project and moving parts of it around distract you from the tasks of creating a research design (at least not for too long).

ℵ SUMMARY

Before beginning your project, you must give careful consideration to design, including how research strategies will be paced and how the method you choose will answer your research question. Consider how you will find participants and what scope for the project will be obtained with your sample. Does your design account for the purpose of the study? How will you locate your study methodologically? What data will you gather and how will you handle these data? Which software will you use? Finally, consider *how* you will use the computer for management and analysis of data.

In this chapter, we have explained the need for careful design of a qualitative project and the special requirements of qualitative research design. Qualitative projects usually involve ongoing processes of design as the researcher designs and reviews the scope of the project and the nature of the data required. We have suggested the questions you should ask and the issues you should consider as you prepare a design, as well as the ways in which you can revisit and revise it as you commence your project. We have described the five steps of establishing your purpose, locating the study methodologically, deciding the scope of your inquiry, planning the nature of your data, and then thinking ahead to the goals you wish to achieve. As you plan, anticipate that your study will involve different stages, and allow time for each—conceptualizing, entering the field, creating a data management system, sampling and theoretical sampling, and final analysis.

Throughout this chapter, we have emphasized that you need to see your project in terms of its overall design. We have discussed combinations of qualitative and quantitative designs. In addition, in this chapter you have met the first of a series of sections on qualitative computing. If you wish to learn software skills, go to Appendix 1 and follow the directions to run the tutorial using software for setting up your project.

〰 RE*J*OURCE*J*

Beginning Design

Hart, C. (1998). *Doing a literature review: Releasing the social science research imagination.* London: Sage.

Piantanida, M., & Garman, N. B. (1999). *The qualitative dissertation: A guide for students and faculty.* Thousand Oaks, CA: Corwin Press.

Richards, L. (2005). *Handling qualitative data: A practical guide.* London: Sage.

Resources for Design

Bernard, H. R. (2000). *Social research methods: Qualitative and quantitative approaches.* Thousand Oaks, CA: Sage.

Creswell, J. W. (1994). *Research design: Qualitative and quantitative approaches.* Thousand Oaks, CA: Sage.

Creswell, J. W. (1998). *Qualitative inquiry and research design: Choosing among five traditions.* Thousand Oaks, CA: Sage.

Denzin, N. K., & Lincoln, Y. S. (Eds.). (1998). *The landscape of qualitative research: Theories and issues.* Thousand Oaks, CA: Sage.

Goffman, E. (1989). On fieldwork. *Journal of Contemporary Ethnography, 18,* 123–132.

Lofland, J., & Lofland, L. H. (1995). *Analyzing social settings* (3rd ed.). Belmont, CA: Wadsworth.

Marshall, C., & Rossman, G. B. (1999). *Designing qualitative research* (3rd ed.). Thousand Oaks, CA: Sage.

Mason, J. (1996). *Qualitative researching.* London: Sage.

Maxwell, J. A. (1996). *Qualitative research design: An interactive approach.* Thousand Oaks, CA: Sage.

May, T. (Ed.) (2002). *Qualitative research in action.* Thousand Oaks, CA: Sage.

Miller, D. C. (1991). *Handbook of research design and social measurement* (5th ed.). Newbury Park, CA: Sage.

Morse, J. M. (1991). Approaches to qualitative and quantitative methodological triangulation. *Nursing Research, 40*(2), 120–123.

Morse, J. M. (2002). Principles of mixed and multimethod design. In A. Tashakkori & C. Teddlie (Eds.), *Mixed methodology: Combining qualitative and quantitative approaches*. Thousand Oaks, CA: Sage.

Patton, M. Q. (2002). *Qualitative research & evaluation methods* (3rd ed.). Thousand Oaks, CA: Sage.

Richards, L. (1999b). Qualitative teamwork: Making it work. *Qualitative Health Research, 9*, 7–10.

Richards, L. (2005). *Handling qualitative data: A practical guide*. London: Sage.

Silverman, D. (2005). *Doing qualitative research* (2nd ed.). London: Sage.

Resources Describing Analysis

Coffey, A., & Atkinson, P. (1996). *Making sense of qualitative data*. Thousand Oaks, CA: Sage.

Dey, I. (1995). *Qualitative data analysis: A user-friendly guide for social scientists*. London: Routledge.

Flick, U. (1998). *An introduction to qualitative research*. London: Sage.

Miles, M. B., & Huberman, A. M. (1994). *Qualitative data analysis: An expanded sourcebook* (2nd ed.). Thousand Oaks, CA: Sage.

Sapsford, R., & Jupp, V. (Eds.). (1996). *Data collection and analysis*. London: Sage.

Wolcott, H. F. (1994). *Transforming qualitative data: Description, analysis, and interpretation*. Thousand Oaks, CA: Sage.

Software and Design

Bazeley, P. (1999). The bricoleur with a computer, piecing together qualitative and quantitative data. *Qualitative Health Research, 9*, 279–287.

Bazeley, P. (2003). Computerized data analysis for mixed methods research. In A. Tashakkori & C. Teddlie (Eds.), *Handbook of mixed methods in social and behavioral research* (pp. 385–422). Thousand Oaks, CA: Sage.

Bazeley, P. (in press). The contribution of computer software to integrating qualitative and quantitative data and analysis. *Research in the Schools*.

Bazeley, P. (forthcoming, 2007). *Qualitative analysis with NVivo*. London: Sage.

Bryman, A. (2006). Integrating quantitative and qualitative research: How is it done? *Qualitative Research, 6*(1), 97–113.

Gibbs, G. (2002). *Qualitative data analysis: Explorations with NVivo*. London: Open University Press.

Kelle, U. (Ed.). (1995). *Computer-aided qualitative data analysis*. London: Sage.

Lewins, A.F., & Silver, C. (in press). *Qualitative data analysis software: A step-by-step guide*. London: Sage.

Richards, L. (1998). Closeness to data: The changing goals of qualitative data handling. *Qualitative Health Research, 8*, 319–328.

Tesch, R. (1990). *Qualitative research: Analysis types and software tools*. London: Falmer.

Thompson, P. R. (1998). Sharing and reshaping life stories: Problems and potential in archiving research narratives. In M. Chamberlain & P. R. Thompson (Eds.), *Narrative and genre* (pp. 167–181). London: Routledge.

Weitzman, E., & Miles, M. B. (1995). *Computer programs for qualitative data analysis*. Thousand Oaks, CA: Sage.

Part II

INSIDE ANALYSIS

5

Making Data

One of our goals in the preceding chapters has been to demystify the process of getting started on a qualitative research project and to show it as a process of informed choice, reflection, flexible planning, and decision making. In this chapter, we take you through these specific processes. As you are preparing to start a project, it is important that you know what these processes will be like and have a sense of the kinds of data you need to make in order to attain the kinds of outcomes you have planned.

In social processes, it helps to think of qualitative data as *made* rather than merely "collected." To speak of data as being "gathered" or "collected" is to imply that data preexist, ready to be picked like apples from a tree. Gathering apples from a tree changes the context of the apples (they are in a basket instead of on a branch) but makes no inherent change in the apples themselves. This is not so with data. Qualitative researchers collect not actual events but representations, usually reports or accounts of events. Talking of "collecting" data denies the agency of the researcher. Try challenging this convention by referring to data as being "made." *Making data* is a collaborative, ongoing process in which data are interactively negotiated by the researcher and participants; the data are rarely fixed and unchanging, never exactly replicating what is being studied. And of course, like any collaborative process, making data is complex and, in the laboratory/experimental sense, impossible to control.

All of these observations are true of research employing questionnaires as well as qualitative interviews. The stereotypical pollster with a foot in the door and a clipboard-mounted checklist instrument is probably more active than the participant observer in defining a situation in which certain answers are more or less likely, but qualitative data have particularly intriguing relationships with their researchers, the participants, and the realities they represent.

≋ WHAT ARE DATA (AND WHAT ARE NOT)?

The researcher, in very many ways, selects from among what he or she has seen or heard and then decides what elements, from all this input, are and are not data. When observing a setting, for example, a researcher will focus on some things and pay less attention to others—those things considered unimportant, irrelevant "noise"—excluding them from data collections or from analysis. Those things that the researcher considers interesting and relevant to the research topic or question catch his or her attention and are recorded in field notes or on audio- or videotape. This focusing is not bias—it is simply the way fieldwork progresses; not everything is relevant, and directing one's gaze toward significant events permits one to move along without getting bogged down in everyday trivia.

The process of recording compartmentalizes individual incidents. A researcher views a scene *interpretively*, responding from personal beliefs and values and filtering what he or she sees or records. Some actions will be implicit and so taken for granted that the researcher will not even record them. Other things will jar the researcher or appear surprising, and he or she will give them greater weight in the data, perhaps so much so that they overshadow other activities. Some things will distract or bore the researcher; others will excite or surprise him or her. Thus the researcher literally makes data.

It is a mistake to assume that technology removes interpretation. Videotaping, for example, delimits an incident to what the researcher chooses to tape and the record to whatever can be captured within the frame itself, giving each taped incident a beginning and an end. Writing, with the limitations of language, memory, and the slow speed at which one can write, makes a summary representation of the actual scene.

The Researcher in the Data

Data are created in a particular form according to the method used. The research question determines the nature or type of research context the researcher must select and also indicates the type of participants and the form of the data (e.g., observational, verbal, or visual). These requirements delimit the researcher's choice of a research site and also indicate the nature of the resulting data.

Within this context, researchers make data in collaboration with their participants. The same technique—interviewing, for example—can make very different data records given different settings, research goals, and relationships between the researcher and those studied. The researcher may have an objective relationship with participants or a full and equal relationship, in which the participants are considered to be coresearchers. Compare the two projects represented in the data on the CD that accompanies this book. In the "Body Image" project, data were created in interaction with the school-age respondents in "artificial" interview and focus group situations where interviewers led the discussion and prompted the often unwilling or inarticulate participants. The "100 Families" life histories were also gathered in interviews, but those situations were much less directive, so the documents flow as natural narratives.

Making data is not a passive process; rather, it is a cognitive process that requires tremendous investment on the part of the researcher. It requires extraordinary concentration, and it is the nature of the questions asked and the attention that the researcher gives to the participants and to detail that determine the quality of the data collected. For example, when the researcher is assuming the listening role in an interview, the participant is constantly observing the researcher and picking up nonverbal cues that implicitly guide the interview. Thus the researcher must actively explore and unravel all aspects of the research questions and the interview context.

Good Data/Bad Data

Given the varieties of processes involved in making data, the interaction between the researcher and the participant, and the collaboration necessary in data collection, data inevitably vary in quality. What constitutes "good" data for a project also varies, of course, with the research goals and opportunities. The collection of good data requires the best possible collaboration with participants. The researcher must find the right way to share with the participants what he or she is studying—not only for ethical reasons but so the participants may trust and help the researcher.

Within this variety, experienced researchers have a sense of what constitutes good data. The records are rich, thick, and dense, offering enough detail to allow someone to comprehend the situation or understand the setting without asking additional questions. Good data are focused, but not focused so tightly that the context is omitted from or restricted within the description. Good data have density. For example, in an interview project there is some repetition between various participants so that one interview, at least in part, confirms or builds on other interviews. Good data are developed in careful recognition of participants' perspectives. An interviewer, for example, guides interviews without "leading" the participants. The participants must always have enough space to present their own perspectives and have their say.

From this description of good data, we can learn what constitutes data that are bad or not helpful. Again, generalizations are dangerous: What has no use for one research purpose may be valuable for another. Some data may disappoint, for example, because the researcher is too "present" in the data. In an interview project, if the interviewer asks leading questions or interrupts the participant, the data will explain things out of order, in a disjointed manner, rather than present the participants' story sequentially. In observational research, a researcher can easily become an artificial presence in a setting. Ideally, researchers have to maintain a delicate balance in which they are both in (a part of the scene) and out (not dominant). Researchers develop the skill of building this "in-but-not-in" relationship over time, as they develop the ability to create appropriate relationships with participants.

It is important to note that most projects will include the making of some data that are off the topic or that prove irrelevant, but this does not mean they are bad data. Although a research question focuses the research study, the researcher may not always be aware of all the factors that are important or unimportant at the beginning of the study. It is prudent for the researcher to keep the data broadly focused at the beginning of the study so that he or she will be sure to collect all the necessary contextual data. Later, when the research becomes more focused, the researcher will find that some of the data from the beginning of the study are not pertinent. Researchers usually retain such data (sometimes called *dross*), as they may prove relevant later to emerging theoretical schemes, or they may be used for secondary analysis.

❃ WAYS OF MAKING DATA

How do you know what ways of making data are appropriate to your study? Data in qualitative research take many forms, and these do not belong to any particular methods (see Table 5.1). Almost all forms of

Table 5.1	Techniques for Making Qualitative Data	
Technique	*Characteristics*	*Used Commonly In*
Unstructured, interactive interviews	Relatively few prepared questions; may be only one or more grand tour questions. Researcher listens to and learns from participant. Unplanned, unanticipated questions may be used; also probes for clarification.	Ethnography, discourse analysis, grounded theory, narrative inquiry, life history, case study
Informal conversations	Researcher assumes a more active role than in interactive interviews.	Phenomenology, ethnography, grounded theory
Semistructured interviews	Open-ended questions are developed in advance, along with prepared probes. Unplanned, unanticipated probes may also be used.	May be used in ethnography and grounded theory, or as a "stand-alone method"
Group interviews	Tape-recorded or videotaped; 6–8 open-ended questions are asked. Facilitator stimulates dialogue among participants.	Focus groups (a particular type of group interview) used across methods; informal groups may be used in ethnography
Observations	Field notes may be recorded as notes (and later expanded) or into a Dictaphone and later transcribed. Participant or nonparticipant observation (dependent on the extent to which the researcher participates) may be used.	Ethnography, grounded theory, supplement to interviews in all methods
Videotapes	May be retained whole for replaying and reviewing or summarized or transcribed (optionally with illustrations retained).	Ethnography, ethology

(Continued)

Table 5.1 (Continued)

Technique	Characteristics	Used Commonly In
Photographs	May be used to illustrate and facilitate recall.	Many methods, especially ethnography
Maps	May be stored and referenced.	All methods where understanding a site is important
Documents	May be collected during project and used to give background or detail.	All methods
Diaries, letters	May be retained and studied in detail or summarized.	Many methods, especially life history
Indirect methods of representing	Researcher finds ways of simulating or representing the phenomenon studied.	All methods

qualitative inquiry may at some stage use some type of interview strategy, and most other ways of making data are also shared across methods. These include recording dialogues as they occur in the setting, making recordings or photographs as one observes the setting, writing field notes, recording descriptions of the setting, and collecting documents or diaries about the setting. Other indirect modes of learning about what goes on in the setting studied include simulated observation, through exploration of literature and other media, such as movies.

Our advice is that you read widely about the skills each technique requires, the ways they can be combined, and choices between types of interview and other strategies. Below, we briefly sketch some ways of making data. For each, we discuss where it is appropriate (or not) and what issues you need to consider in its application.

Interviews

Most researchers approaching qualitative research expect that they will conduct interviews. But interviews may not be the most appropriate way of making data and often will not be the easiest. The types of interview used and the ways in which the interviews are conducted depend on the research question and method (see, e.g., Holstein & Gubrium, 2003; Kvale, 1996).

Unstructured, Interactive Interviews

The most common type of qualitative interview is unstructured and interactive. Conducted well, this type of interview offers a participant an opportunity to tell his or her story with minimal interruption by the researcher. Unstructured interviews are most appropriately used in studies where the researcher seeks to learn primarily from respondents what matters or how procedures are understood.

When you are preparing to conduct an unstructured interview, you should select a quiet, comfortable, and private setting. You must let the participant know that the interview will take at least 2 hours. If the interview is being conducted in the participant's home, make certain that the television, radio, and telephone are switched off; the sound of TV or radio, even in the background, will distract you and the participant and will harm the quality of the tape recording of the interview, and incoming phone calls may interrupt the participant. Unless the research design demands that interviews be conducted jointly by coresearchers, you should be alone with the participant while you conduct the interview.

How much preparation should you do prior to conducting an interview? Some researchers prepare approximately six to eight broad, open-ended questions. However, as the purpose of the interview is to elicit the participant's story, asking such questions initially may structure the interview and render it invalid. It is better if you simply let the participant tell his or her story, and then, if you have not learned about all aspects of whatever it is you want to know, you can ask questions when the participant finishes speaking, or at a second interview. Your role is to listen and let the participant tell his or her story without interruption.

Beware of the assumption that any good listener can interview. Skill is required to generate the needed narrative without interrupting the respondent's story. This is not an easy task. The role of the researcher as an active listener cannot be overemphasized. Like any good listener, a skilled interviewer is responsive and interested. Guidance of the interview ideally should take the form of your helping to maintain the conversation rather than interrupting the participant's flow of thought. Probes or questions designed to elicit further information may sometimes be developed ahead of time; they may also be unplanned, arising from the context of the interview. But you need to take care that you offer such input only as appropriate during the interview; often, it is best left until the end.

Before the interview, the participant already knows the general subject of the research, as the researcher must give him or her this information during the process of obtaining consent for the interview. Often, however, the researcher does not plan the interview questions in advance, or may prepare only an opening question (sometimes called a "grand tour" question). The grand tour question is intended to focus the participant on the topic, and the researcher, in essence, assumes a listening stance. During the interview, the researcher may ask an occasional question for clarification but generally permits the participant to tell his or her story without interruption.

Semistructured Questionnaires

Sometimes, the researcher knows enough about the phenomenon or the domain of inquiry to develop questions about the topic in advance of interviewing, but not enough to be able to anticipate the answers. When this is the case, the researcher can prepare and trial-run a semistructured questionnaire. For such a questionnaire, the researcher designs open-ended questions, arranged in a reasonably logical order, to cover the ground required. Usually, the interviewer will ask the same questions of all of the participants, although not necessarily in the same order, supplementing the main questions with either planned or unplanned probes. The interviews are normally tape-recorded and transcribed in preparation for analysis.

The use of semistructured interviews is appropriate when the researcher knows enough about the study topic to frame the needed discussion in advance, as in the "Body Image" study. Such interviews offer the researcher the organization and comfort of preplanned questions, but also the challenge of presenting them to participants in such a way as to invite detailed, complex answers.

Sometimes, if participants cannot be interviewed or if the topic is sensitive or embarrassing, the researcher may ask participants to write their answers on the semistructured questionnaire itself; the instrument is formatted with spaces after each question in which participants can write their open-ended responses. These responses are usually transcribed into the computer in preparation for analysis.

Conversations

Researchers who are interested in linguistic or discourse analysis often record dialogue and analyze it as it occurs (for instance, the dialogue between a doctor and a patient). There is normally little or no interjection from the researcher in these data, and the data are usually tape-recorded. If dialogue is your focus, this strategy is, of course, necessary and relatively straightforward. But beware of recording dialogue that may not address your topic.

Group Interviews

Group discussions and more-guided group interviews can take many forms. These include unguided conversations, formal meeting interactions, social gatherings, and multiple-respondent interviews. Anthropologists encounter or construct such opportunities throughout their studies, recording what they learn in field notes. Such group interview situations are to be distinguished from focus groups, which consist of individuals brought together by the researcher to focus specifically on one, usually narrow, topic. Often researchers use focus groups to gain understanding of the research domain relatively quickly. They may also employ focus groups to scope a project early in its design; such groups can provide researchers with information about a topic's dimensions or people's attitudes on an issue.

Focus group research has become very popular in many areas; it has its own literature, some of which specifies rules and requirements very narrowly (see the list of resources at the end of this chapter). Most authors who offer advice about conducting focus groups assert that a group should normally consist of about 6–10 participants and that a typical group session lasts from 1½ to 2 hours. The group facilitator is responsible for the quality of the data; he or she introduces the questions and ensures that the conversation is balanced (e.g., not dominated by one or two participants) and that the dialogue stays more or less on topic and does not get stuck on one point for too long. Because the strategies are so specifically detailed in the literature, conducting focus groups may sound easy. If you are considering using focus groups in your research, we advise you to think carefully about the sort of information likely to be obtained through this method, as well as the context of that information. And don't underestimate the skills required of a focus group facilitator.

Observations

Observing is the most natural of all ways of making data, but observing unobtrusively is extremely difficult. Researchers may be able to gain an understanding of some behaviors only through observation: It may not be possible to interview some participants (due to language differences or other inability to communicate), or participants themselves may not be aware of some of their behaviors. The observer's perspective may be different from that of the participants, or there may be a very weak link between reported behavior and actual behavior, but observation is a primary strategy of making data even in situations where interviewing would be easy. The assumption behind most observational strategies is that they enable the researcher to learn what is taken for granted in a situation and to discover what is going on best by watching and listening.

Observational techniques differ primarily in the visibility and involvement of the researcher in the setting. The usual distinction between *participant observation* and *nonparticipant observation* hides the fact that there are myriad ways of watching and listening. No observer is entirely a participant, and it is impossible to observe in almost every nonexperimental situation without some participation. If you plan to undertake observation, we advise you to read not only texts about it but also monograph accounts of the experience and the resulting data (for many relevant references, see Atkinson & Hammersley, 1994).

Traditionally, researchers have recorded their observational data, along with their own interpretations, in the form of field notes. The researcher observes a setting for a short period of time and retreats to a quiet place to record and reflect on the observations. Often, researchers supplement their field notes with maps or diagrams that they use to record interaction patterns or to clarify the positions of features within the setting. Occasionally, they include in their observations records of the actual words of participants' conversations in the setting. Researchers also commonly use photographs to support their observational data, analyzing them to gain additional insights into the setting.

Currently, videotaping is becoming more widely used in observational settings; as the use of video cameras is becoming increasingly common, the presence of these cameras seems less intrusive to participants, and fears among researchers that they change behavior are now diminishing. But two warnings are relevant: First, if you videotape, practical and ethical considerations remain and must be addressed (Will taping change participants' behavior and/or inflict stress or embarrassment?); and second, it is important that you avoid the trap of assuming that by having taped an event you have interpreted it. Videotaping does not replace the process of compiling field notes—data assessment and interpretation. Beware of the study design that produces hours of uninterpreted tapes with no provision for how and when they will be observed and analyzed.

Photography

Researchers may use photography as an independent way of making data to record a setting or scene, to record directly "how much" or "how many," or to provide illustrations. Researchers' photographs may serve as data in historical studies or in life history studies, or they may provide background as a part of the ongoing data collection scene. Sometimes, researchers retrieve photographs from other collections, with the aim of providing a means for comparison, to illustrate change. With the availability of computer storage, photographs have become more easily accessible, and researchers are increasingly integrating photography into their data in innovative ways. For example, Westphal (2000), in her study of the use of urban space, gave local children disposable cameras and asked them to take photos of places of significance to them. Westphal's interviews with these young respondents, who had little education, were facilitated by discussion of their photos.

Documents

Data often consist of documents that exist independent of the research process. These may be institutional records, such as school or management records, policy statements, or home records. Participants' diaries or letters also provide many insights into participants' lives and may be used as data. Researchers gather these data and perhaps copy or scan them for direct insertion into the data file.

Indirect Strategies

Researchers have employed indirect data sources fruitfully in many different methods to generate interpretations and responses to hypothetical situations. Finch (1984) used *vignettes* to discuss the highly emotive topic of family obligation for caring and found that participants could respond firmly about the duties or problems of vignette characters even when they were unwilling to discuss their own. *Simulated observation* involves using participants as actors to show the researcher what they normally do. For example, in her study of traditional birth attendants in Fiji, Morse (1989a) asked the participants to show her the positions the women they attended normally used while giving birth.

Indirect data that may illuminate or provide insights into participants' experiences and responses can be found in literature, movies, theater, and art. These sources provide interpretations of real life that researchers can use in analyzing dominant discourses or ideologies, historically situated interpretations, or background clues as to what is going on in more direct data.

WHO MAKES DATA?

In quantitative inquiry, it is very common for research assistants to help investigators in "collecting" data, usually through the use of precoded research instruments. In qualitative inquiry, however, where researchers themselves are more active in making data, the use of research assistants is, if not less common, certainly less often regarded as unproblematic. Principal investigators who collect their own data argue that by doing so they ensure that the data are of excellent quality and that the interpretive process (such as field note notations about "what is going on") and the possible effect of the researcher on the setting are sufficiently recorded. At the same time, researchers commonly work on projects in loosely or tightly coordinated teams.

Some qualitative researchers do use research assistants to collect data. Indeed, Glaser (1978) has asserted that researchers' conducting interviews by themselves is "a waste of time." His approach is to interview the interviewers to find out what is going on in the data. There is agreement among researchers, however, that when first-level coding is used, the principal investigator should at least do the bulk of the coding.

Conducting an interview or observing in a setting is a task that requires extraordinary interpersonal skills, a firm focus on the project's purposes, and solid theoretical knowledge (because data are representations that need to be interpreted, they are interpreted, and the process of recording does not capture everything). If you decide to hire research assistants to help with data collection, be certain that they are theory-smart and method-wise in their work. More important, ensure that you work with them as a team, and that you stay in constant communication with them to discuss interpretations and discoveries.

If your project requires teamwork or collaboration with other projects, build this into the research design; the need for team research should inform your choice of strategies for making and recording data as well as your planning of time and budget.

⟩⟩⟩ TRANSFORMING DATA

Preparing data for analysis is a process of transformation. Each research event is transformed from an actual happening to a form that can be handled and manipulated in the process of analysis. Ideally, this process keeps the data as close to the actual events as possible. It may be achieved through the reduction of data in the form of detailed field notes or recordings in other media (audiotape, videotape, or photography) to text (transcription or a summary description). Be aware of how massive this reduction is and how interpretive the process.

Thus data ready for analysis are several steps removed from the actual event. The researcher must always remember to interpret data in this context. For example, interviews of participants about their experiences or perceptions of an event should be judged not in terms of the accuracy of the participants' recall of the actual event but in terms of the accuracy of their recall of *how they felt or experienced or perceived the event at the time.*

As Table 5.1 shows, there are limited numbers of ways of recording data used across many strategies of making data. None is inherently superior to any other. Some researchers who are trained as observers assert that the best records come from listening and remembering, with no technology intervening. Others see video recordings as the richest and most lasting records of any given scene (or rather of what was observed), because videotapes allow the researcher to review, examine, and reexamine the scene by replaying the action slowly, even frame by frame. This may assist the researcher in obtaining a more accurate description of the action, particularly if the movement is very fast or if more than one person must be observed at a time. But like all data gathering, reviewing videotapes is an interpretive process. Just as "objective" field notes record what the researcher observed, *not what happened,* a video captures only what was recordable and what was taped.

How does the recording process influence the reliability and validity of a study? The answer depends on the research question and the type of detail and data needed to answer the question. Strategies for data collection must enable the researcher to gather data at the level needed to answer the questions. If a question addresses types of touch, and touch is silent and transient, then videotaped data are required. If the question addresses the detailed content of conversation, then audiotaped dialogues will be needed. If the question addresses relationships, then field notes or taped interviews may be used, and so forth. In other words, different sorts of interpretive records are relevant for different projects, but all records may be interpreted.

Traditional notions of reliability and validity involve ensuring that the processes of data creation and interpretation record the phenomena of interest as closely as possible and that two researchers working independently obtain data that are as similar as possible. However, if one considers the process of analysis to be *interpretive,* the replication of events is less relevant than a recording that provides insight into exactly what is "going on."

＊ MANAGING DATA

The volume of data that qualitative researchers must manage is enormous. Researchers tell stories of "drowning in data," of stacking piles of data in their basements, or of not being able to use their dining room tables for several months while the process of analysis was ongoing. There are two aspects to this problem—how to manage the amount of data and how to manage the data *records.*

What amount of data is sufficient? The answer is in the previous chapter. As you create and assess your research design, you will ask this question regularly. The answer will become clearer during the simultaneous process of making and analyzing data. Theoretical sampling provides *just enough* data. Look for indicators of *saturation* (the replication of data or the verification of incidents/features/facts by several participants) and confidence that adequate data have been obtained. Thus, in a well-designed and well-conducted study, data are not overcollected, but they are well managed.

How do you manage the data records? This involves physical handling of the growing heaps of records as well as intellectual handling of their growing complexity. Physical handling may seem easy, but it requires considerable self-discipline. Clerical duties abound: Items must be clearly labeled and easily located. Where anonymity of those you are studying is required (and this is the usual situation), it is important that you create and consistently use a system of pseudonyms or codes; if you are using and transcribing tapes, you should place these identifier codes on the tapes and in the headers of the transcriptions. If necessary, you may keep, in a secure place, a key that links names with code numbers. Normally, you will want to destroy this list as soon as possible. You should store all audiotapes and videotapes in a locked cabinet and take care not to expose the tapes to extreme temperatures, static electricity, or magnetic fields. As the data are transcribed, you should back up the original files on another safe medium. Note that floppy disks are *not* a safe storage medium—use zip disks, CD-ROMs, or another hard disk. As you work with the data, you will, of course, need to back up your growing and changing project.

The computer offers a great deal of help in the physical management of records and the fluid management of ideas. Researchers now normally store and manage data transcripts, written field notes, and other text documents on the computer, whether or not they use any specialized software. Programs designed for qualitative research offer a very wide range of ways of managing data using folders, document names, definitions, memos about documents, and coding. They also offer researchers new ways of storing their impressions and ideas from the earliest stages of their research so that they can easily locate them again. (See the final section of this chapter, and, for directions to tutorials, Appendix 1.)

Whatever strategies for analysis researchers use, they will always want access back to the original documents. So data must be located to facilitate easy and quick retrieval. Most retrieval systems allow researchers to gain access using document identifiers, participant pseudonyms, and, as the coding analysis proceeds or develops, codes or categories.

The challenges of data management for the data your project needs, and the sort of access you will require to your data records, should be considered early in your project.

Managing Focus Group Data

These data management issues are clearly shown in the challenges of handling data from focus groups. When records of focus groups are not carefully managed, reports are often very simplistic assertions about what the group "concluded" or "felt." Such reports waste the data about process and context that made the group interesting.

Unlike an interview, a focus group provides a record created in discussion between participants. Whether this would be an advantage to your project will depend on the research area, the question, and the method you are using. For example, if you are entering a previously unresearched field, groups may be the best way to discover the issues or perceptions that will matter to your project. If the topic is complex, focus groups may help you to distinguish between the dimensions that matter to participants.

It is important to recognize that focus groups take skill and that the data they provide can be very difficult to analyze. One of the myths of focus group research is that focus groups are a quick and cheap way of "getting" a lot of data from a lot of people (Morgan, 1993). While any group can make a lot of words, it will require skilled moderation to ensure that these are useful to your project, and then skilled data management to use them. If this method appears to fit with your project design, prepare for them carefully, reading about designing and moderating groups, and managing the data that result. Focus groups are used very differently in different research settings, and you should be clear about the methods developed in yours.

When you tape even a short focus group, it transcribes to a *long* record. Even if it did not seem to be a very challenging group discussion, you will find you have a record of complex interaction. Reading a transcript, you will immediately want access to information about its context. Who said that? (Was it a woman or a man?) Were they serious in this comment or laughing? (Do we need the voice record, or the video image, to check?) What earlier comment from another participant prompted their response? (And who was that?) Did that earlier remark make them change their minds? (What had they said before on this topic?)

How do you manage such complex data? When confronted by complex, unstructured record, a good first step is to look for *structure*. A well-designed focus group is usually structured by the moderator's input and the participants' interaction. Using computer software, you can automatically gather from all your focus groups all the answers to each of the moderator's questions, so they can be read together. Then you can gather everything said by each participant in a group, so you can follow their contributions throughout the discussion.

For a second step, look for *context* and how to manage access to it. With focus group data, there are two sorts of context available to you. You have the original tape, which records much more context than the transcript, and from which you could learn, for example, whether that participant was laughing and who was talking behind him. You may wish to plan to use software to code the audio- or videotape, or to place links to "clips" of tape-recording the important moments. The second source of context is the information about participants' characteristics, usually gathered as a group is recruited. If you have gathered and coded every-thing each respondent said, you can store the relevant information about their characteristics with that coded data. Now it is possible to ask what the women said on a particular topic and to explore whether they had previously held a different view.

\\\\ THE ROLE OF DATA

The aim of data management in qualitative research is not merely to pro-tect the researcher from overload or data wastage. As you develop a data management system, you will discover that well-managed data inform, even lead, the process of inquiry. When you maintain a balance between data management and data direction, the role of the data can be pivotal, and the process exciting.

Data lead inquiry in three ways. First, they may demand to be treated by a particular method, and no amount of forcing will permit you to mold the data to fit the originally planned question and method (e.g., you intended to conduct structured interviews, but the respondents talked in stories that demand narrative analysis). Don't force. Data will be wasted if you do not treat them within the most appropriate method.

Second, data may direct you to a new way of working in the next stage of a project. For example, you may discover that you don't understand the processes the participants are discussing in interviews. The interview data are directing you toward a stage of observation. Thus, in her study of outer suburbia, Richards (1990) initially relied on surveys and inter-viewing, but the data increasingly "directed" her toward observation and detailed analysis of discourse. These strategies provided her with the key to understanding the participants' interlocking ideologies and helped her to make sense of interview answers that had appeared to be contradictory.

Third, data may tell you what you can't do. In her suburban study, Richards (1990) discovered that data on interactions among neighbors were not available through direct interviewing in an area where residents were ideologically required to express enthusiasm about their interactions with neighbors. Observation provided quite different results. In such a situation, the researcher must rework the research design, and the report must account for this change of plans.

Expect, then, to be engaged in a dialogue with your data. Recognizing the ways in which data direct the adjustment of research strategies (and sometimes even of research question and study purpose) places the data in a dominant position, one that drives the study. Such realization is sometimes painful, but forcing data into inappropriate studies is more painful and unproductive. This occurs most often when a researcher begins by preparing a proposal while "wed" to a research approach that fits poorly with the topic. The researcher risks making data inappropriate for the method and later may find it impossible to mold the data to the chosen method. Often the only way to proceed is to use the method demanded by the data.

Data can play a definitive role in determining whether a question can be answered. Research into the phenomenon of intuition in nursing has sought data on how nurses can predict a change in a patient's condition before the physiological changes occur and show on the patient's monitor. Nurses are unable to describe the look that patients get, yet researchers have conducted much of their research by interviewing nurses about what they intuit rather than by observing patients to learn the behavioral signs they manifest (Morse, Miles, Clark, & Doberneck, 1994).

Sometimes the data required by a question and method may not be obtainable. At the extreme, one cannot interview infants or aphasic adults; participants with Alzheimer's disease or mental illness often cannot report on their experiences. Participants cannot report on experiences that occurred when they were unconscious, and they often forget or cannot recall their experiences in agonizing pain or other circumstances. Lorencz (1992) explored the experience of being discharged from a psychiatric hospital by interviewing schizophrenics. As an experienced psychiatric nurse, she discovered she was competing with the patients' multiple voices and multiple realities and often waited more than a minute for participants to respond to her questions. When they did answer, their responses often appeared irrelevant, as though they were responding to other, internal voices; thus the interviews were often nonsensical, slow, and convoluted, but they offered glimpses of the patients' realities. Studies with less extreme conditions often also have major gaps in their data: An important event is missed, it proves impossible to find or gain agreement from crucial participants, or those participants most needed are those most elusive.

Sometimes data may be compromised because of the context, such as when the researcher is denied access for political or ethical reasons and adapts to working with the best available data. But even imperfect data can be amazingly interesting and can produce quite satisfactory results. Morse (1992a) found that nurses in trauma rooms insisted they "did not have time to comfort patients" and that the study should be conducted elsewhere. Nonparticipant observation enabled Morse to identify the major types of comfort, but in this setting, recording the confusion—or sorting out the confusion—using participant observation or a stenographer's record proved impossible. Eventually, at another hospital in another country, she and her colleagues received permission to videotape, using cameras attached to the wall—"shooting blind," with poor sound quality and the requirement that the patient's face be obscured. But even these poor data allowed the documentation of the linguistic pattern of speech used by nurses, later dubbed the Comfort Talk Register (Proctor, Morse, & Khonsari, 1996), and of the ways nurses comfort patients (Morse & Proctor, 1998).

⟩⟩⟩ YOURſELF Aſ DATA

Should the researcher be seen as an active participant in the setting studied? It was strenuously argued in early anthropology that researchers should separate themselves from their topics and the people they studied to avoid having their own personal agendas drive the research problem. It was feared that personal involvement made the topic more stressful, that researchers would lose their objectivity, and that the reporting would lose its fairness. Anthropologists argued that researchers could not see a particular culture's values and beliefs if they were immersed in that culture. These objections have now by and large been relaxed, but, as Lipson has noted, this does not mean it is easy to place yourself in your data:

> It takes a real effort to figure out what is you and what is not you. I think it can be done successfully, but I think it takes a lot of experience and a lot of hard work to get there. And a lot of exposure and a lot of self-explorations to find out where your own values are coming from, what your own behavior is, what you're not seeing, and I think it is a very sticky proposition. ("Dialogue," 1991, p. 72; see also Lipson, 1991)

Whether your experience will affect the research is not the question—it will. For all researchers, the big question is how to place that experience. How do you monitor and account for the ways your values, beliefs, culture, and even physical limitations affect the process and quality of data? Two issues are particularly urgent at this early stage: (1) What should your relationship be with those you study? and (2) What role should you and your experience have in the study? As you read the literature where your project is located, you will find inconsistent answers to both questions.

You and Those You Study

Qualitative methods do not usually advocate depersonalizing those studied and denying the researcher's effects on their behavior. However, you may find that the use of the passive voice in reporting ("An inquiry was made to the group" rather than "I asked the group") and terms such as *informants* and *actors* indicate attempts to depersonalize relationships. (We are wary of these techniques, and we agree that it is best to acknowledge and discuss the researcher's agency rather than to elide or deny it.)

Few methods, however, go to the extreme alternative of seeing researchers as a part of the setting, with no constraints on the ways they may influence it, with their interpretation being just one among many, and with those in the setting regarded as not merely participants but coresearchers. On these issues, the feminist qualitative literature has led an uneven debate since Oakley's (1981) enthusiastic writing about interviewing women and Finch's (1984) more cautious concerns about manipulation of the vulnerable. Most recently, these issues have been expressed in literature on the feminist narrative techniques of memory work (Haug, 1987; for classic and recent references, see Olesen, 2000).

Regardless of the point on this range at which your project is positioned, you must consciously choose, negotiate, and maintain the relationship you have with those you study, and you must discuss that relationship throughout your report.

Your Experience as Data

What role should the researcher's personal experience have in the study? When experience brings a particular research problem to the fore, it will drive the study. The researcher's experience may provide a puzzle that leads to a research question. Glaser and Strauss (1968) began their collaborative research on death and dying because both men had experienced the death of a parent in a hospital and both had noted similar features of the experience that were dissatisfying, unexpected, and, in retrospect, "odd." In such a case, the researcher must decide whether or not to disclose his or her personal interest in the topic (which will make the researcher's experience public). Whether the researcher discloses or not, his or her experience must be involved in the study.

There are two approaches you can use to incorporate your own experience in a study. Be forewarned: Both have serious risks and must be done well. The first is to separate your experience from that of others in the study, thus introducing it but segregating it. In this way, Malacrida (1998) explored the experience of having a stillborn child and Karp (1996) investigated the experience of depression. The experiences of others, presented in the main study, validated the researcher's own experiences. Talking to others in the course of inquiry about an aspect of his or her own experience has an important psychological impact on the researcher. The research makes the researcher's own responses and experiences, which had seemed extraordinary to him or her, more everyday, more *normal*. But like all aspects of research, the use and reporting of personal experience must be purposeful. The risk of this approach is that it can easily become self-indulgent. Keep asking yourself *why* your own experience matters to *this study*. Why should you tell the reader about it?

The second way to use your own experience in inquiry is to use it *as data*. You may present your experience as intimately richer and more valid than the reported, secondhand experiences obtained by interviewing or observing others. You can obtain any validation that may be necessary by linking your own experience with the experience of others reported in the literature or with concepts that are already well developed in the social sciences. Thus Arthur Frank (1991) reports on his experience with cancer in his book *At the Will of the Body*.

If you are bringing your own experience into the study, you should go first to the growing and important body of literature on autobiography and autoethnography. Recent works include substantial lists of readings to lead you not only to the excitement of analyzing yourself as data but also to the perils of doing it badly (see, e.g., Ellis & Bochner, 2000).

〰 USING YOUR SOFTWARE FOR MANAGING DATA

Chapter 4 introduced ways in which you can use the computer for creating a project to contain your data and ideas, and immediately for handling the preliminary material that leads to research topic and question. This material is data, and like all the data records you will later create, it must be very carefully and rigorously handled.

As qualitative data records are created, any researcher is challenged to manage their complexity and richness well and responsibly. These are tasks for which software is obviously an aid. Qualitative packages specialize in ways of creating, importing, handling, and managing data records on the computer.

Approaches

Almost all software designed for use by qualitative researchers will handle text. (Some still require that it be plain text. Most handle "rich text" and/or word processor formats.) Importantly, programs have different ways of including or linking to nontext records such as audiotapes or videotapes. If you wish to use such records as primary data, without reducing them to representations, explore software that will allow you to retain them whole and to code "streaming" tape.

All qualitative software will have some ways to store your reflections on the data, your early ideas as they happen, ensuring that these impressions will not be lost and that you can revisit your account of your data making as the project grows.

Explore your software to get a feel for the processes of making data records and handling them on the computer.

Your software should support all the following processes, which you need to learn:

⧵ How to create and edit documents in your qualitative software program, instead of in a word processor

⧵ How to import or link to rich text files so they can be coded and analyzed in your project

⧵ How to store information about the data records and cases they represent (e.g., demographic data)

⧵ How to add annotations and edit

⧵ How to link between documents or parts of data records

⧵ How to handle nontext records so they can be integrated with text

⧵ How to store your reflections in memos

If you wish to try doing it in software, go to Appendix 1, where there is a guide to tutorials available online.

Advances

The list of software functions above includes much that could not be done before computers. When documents were on paper, storing them and marking them up was awkward and sometimes very time-consuming, but it could be done. On the other hand, changing them, finely annotating them and accessing those comments, storing information about them, and linking that data with statistical analysis were often practically impossible.

Methods change with technology, and these lists offer the first glimpse of the effects of software on qualitative method. You can do much more with your data once they are on the computer, and, of course, you can do it with much more data.

This does not mean you should!

Alerts

Four cautions apply here, and they all concern researchers' tendencies to try to fit research to computer programs rather than making the programs work for their projects.

Data types, the volume of data, and the data's heterogeneity should be driven by the research goals and method, not by the computer program.

※ Be very careful not to skew your project to what the computer seems to want. If your program won't handle the sort of data your project requires, devise a new strategy using the program's tools (and tell other researchers about it) or move to another program.

※ Don't create bulk data records just because the computer can handle them. Most qualitative projects are not best done with large volumes of data, and some are destroyed by data bulk. In the processes of designing your study and fitting your research question to a method and ways of making data, you will have good reasons to predict the volume of data to be created. Always avoid the temptation to make the project impressive by expanding the scale of data, even though this may not expand the scope of the project.

☒ Be discriminating and thoughtful in deciding what are data. There is security in the ability of the computer to store and access all the peripheral and often unexpected material that comes your way—background information, unrelated observations, available documentation, and so forth. But your research design should inform your selection. Never refuse possibly relevant data, but don't assume it must be immediately included in your project.

☒ Beware of tidying up your data for the computer. We have argued above that qualitative research rarely thrives on homogeneous data. Your software does not need homogeneity and, indeed, can do less with homogeneous data than with data sources and types that are varied. Any computer program is better than a human brain and far better than a filing cabinet at managing complexity. And qualitative projects almost always require complex data.

☒ ∫UMMARY

In the preceding chapters, we provided information about the diversity of qualitative research methods. In this chapter, we have presented information about strategies for making interview or observational data that will enable you to meet your research goals and address your research question. Data are not made in isolation but must be linked to plans for analysis. We have also discussed the interactive effects of the nature of your data on the project as a whole.

☒ RE∫OURCE∫

Richards, L. (2005). *Handling qualitative data: A practical guide.* London: Sage.
     Chapters 2 and 3 advise on making good qualitative data records and handling them.

On Focus Groups

Carey, M. A. (Ed.). (1995). Issues and applications of focus groups. *Qualitative Health Research, 5,* 413–524.
Finch, J. (1987). The vignette technique in survey research. *Sociology, 21,* 105–114.

Krueger, R. (1994). *Focus groups: A practical guide for applied research.* Thousand Oaks, CA: Sage.

Krueger, R., & Casey, M. A. (2000). *Focus groups* (3rd ed.). Thousand Oaks, CA: Sage.

Morgan, D. (1993). *Successful focus groups: Advancing the state of the art.* Newbury Park, CA: Sage.

Morgan, D. (1997). *Focus groups as qualitative research.* Thousand Oaks, CA: Sage.

Morgan, D., & Krueger, R. (1997). *The focus group kit.* Thousand Oaks, CA: Sage.

Stewart, D., & Shamdasani, P. (1990). *Focus groups: Theory and practice.* Newbury Park, CA: Sage.

On Interviewing

Fontana, A., & Frey, J. H. (2000). The interview: From structured questions to negotiated text. In N. K. Denzin & Y. S. Lincoln (Eds.), *Handbook of qualitative research* (2nd ed., pp. 645–672). Thousand Oaks, CA: Sage.

Gubrium, J. F., & Holstein, J. A. (Eds.). (2002). *Handbook of interview research: Context and method.* Thousand Oaks, CA: Sage.

Holstein, J. A., & Gubrium, J. F. (2003). *Inside interviewing.* Thousand Oaks, CA: Sage.

Kvale, S. (1996). *InterViews. An introduction to qualitative research interviewing.* Thousand Oaks, CA: Sage.

Melia, K. M. (1997). Producing "plausible stories": Interviewing student nurses. In G. Miller & R. Dingwall (Eds.), *Context and method in qualitative research* (pp. 26–36). London: Sage.

Rubin, H., & Rubin, I. (1995). *Qualitative interviewing: The art of hearing data.* Thousand Oaks, CA: Sage.

Spradley, J. P. (1979). *The ethnographic interview.* New York: Holt, Rinehart & Winston.

On Observation

Angrosino, M. V., & Mays de Pérez, K. A. (2000). Rethinking observation: From method to context. In N. K. Denzin & Y. S. Lincoln (Eds.), *Handbook of qualitative research* (2nd ed., pp. 673–702). Thousand Oaks, CA: Sage.

Atkinson, P., & Hammersley, M. (1994). Ethnography and participant observation. In N. K. Denzin & Y. S. Lincoln (Eds.), *Handbook of qualitative research* (pp. 248–261). Thousand Oaks, CA: Sage.

Ball, M. S., & Smith, G. W. H. (1992). *Analyzing visual data.* Newbury Park, CA: Sage.

Jorgensen, D. L. (1989). *Participant observation: A methodology for human studies.* Newbury Park, CA: Sage.

Spradley, J. P. (1980). *Participant observation.* New York: Holt, Rinehart & Winston.

On Narratives and Life Histories

Atkinson, R. (1997). *The life story interview.* Thousand Oaks, CA: Sage.

Beverley, J. (2000). Testimonio, subalternity, and narrative authority. In N. K. Denzin & Y. S. Lincoln (Eds.), *Handbook of qualitative research* (2nd ed., pp. 555–565). Thousand Oaks, CA: Sage.

Denzin, N. K. (1989). *Interpretive biography.* Newbury Park, CA: Sage.

Ellis, C., & Bochner, A. P. (2000). Autoethnography, personal narrative, reflexivity: Researcher as subject. In N. K. Denzin & Y. S. Lincoln (Eds.), *Handbook of qualitative research* (2nd ed., pp. 733–768). Thousand Oaks, CA: Sage.

Riessman, C. K. (1993). *Narrative analysis.* Newbury Park, CA: Sage.

Tierney, W. G. (2000). Undaunted courage: Life history and the postmodern challenge. In N. K. Denzin & Y. S. Lincoln (Eds.), *Handbook of qualitative research* (2nd ed., pp. 537–553). Thousand Oaks, CA: Sage.

On Videotaped Data

Albrecht, G. L. (1985). Videotape safaris: Entering the field with a camera. *Qualitative Sociology, 8,* 325–344.

Ball, M. S., & Smith, G. W. H. (1992). *Analyzing visual data.* Newbury Park, CA: Sage.

Bottorff, J. L. (1994). Using videotaped recordings in qualitative research. In J. M. Morse (Ed.), *Critical issues in qualitative research methods* (pp. 244–261). Thousand Oaks, CA: Sage.

Couch, C. J. (1986). Questionnaires, naturalistic observations, and recordings. In C. J. Couch, M. A. Katovich, & S. L. Saxton (Eds.), *Studies in symbolic interaction, supplement 2: The Iowa school (Part A)* (pp. 45–59). Greenwich, CT: JAI.

Farber, N. G. (1990). Through the camera's lens: Video as a research tool. In I. Harel (Ed.), *Constructionist learning* (pp. 319–326). Cambridge: MIT Media Laboratory.

Harel, I. (1991). The silent observer and holistic note taker: Using video for documenting a research project. In I. Harel & S. Papert (Eds.), *Constructionism* (pp. 449–464). Norwood, NJ: Ablex.

Lomax, H., & Casey, N. (1998). Recording social life: Reflexivity and video methodology. *Social Research Online, 3*(2). Retrieved May 8, 2001, from http://www.socresonline.org.uk/socresonline/3/2/1.html

6

Coding

Any researcher who wishes to become proficient at doing qualitative analysis must learn to code well and easily. The excellence of the research rests in large part on the excellence of the coding. (Strauss, 1987, p. 27)

I n qualitative research, everyone uses the term *coding*, but different researchers mean many different things when they use that term. Why is this variation so rarely discussed? Indeed, researchers often get into difficulties because they are led to coding without a clear picture of its purpose.

There are many ways of coding and many purposes for coding activities across the different qualitative methods. They all share the goal of getting from unstructured and messy data to ideas about what is going on in the data. All coding techniques have the purpose of allowing the researcher to simplify and focus on some specific characteristics of the data. And all of them assist the researcher in abstracting, or "thinking up," from the data. The common use of the term *coding* obscures crucial differences among techniques in the ways they link ideas and data and also in the relative importance of coding (see Coffey & Atkinson, 1996; see also the debate generated by Coffey, Holbrook, & Atkinson, 1996). Every qualitative researcher, however, has to be able to code. And it is

Authors' Note: The material in this chapter draws on previous work by Lyn Richards written with Tom Richards (L. Richards & T. J. Richards, 1994; T. J. Richards & L. Richards, 1994; Richards & Richards, 1995) and with Pat Bazeley (Bazeley & Richards, 2000).

essential to see that qualitative coding is an entirely different process, with different purposes and outcomes, than coding in quantitative studies. (For a detailed comparison, see Richards, 2005, p. 86.)

In this chapter, we start with the techniques of qualitative coding—what researchers *do* to data on the way to abstraction. We distinguish among three kinds of coding, all of which contribute differently to the processes of analysis. The first is the storage of information, sometimes termed *descriptive coding* (Miles & Huberman, 1994). The second is coding in order to gather material together by *topic*. And third is the coding used when the goal is the development of concepts (*analytic coding*). Finally, we discuss the broader goal of the identification of *themes*.

Novice researchers find *topic coding* the most accessible of these techniques. You will need to gather material by topic if you wish to reflect on all the different ways people discuss particular topics, to seek patterns in their responses, or to develop dimensions of that experience. When a research design clearly addresses specific topics, coding them is an obvious task. By simply collecting all the answers via a coding technique, the researcher can get a new "cut" on the data.

When a researcher does topic coding on paper, the technique usually involves bringing copies of passages physically together. The researcher identifies portions of text as being associated with a particular topic, copies them from the original document, and places them in a labeled topic file. When the researcher wishes to reflect on a given topic, he or she goes to that file. The process works fairly well as long as any given text passage is about only one topic. Like a child sorting marbles by color, the researcher can gather data bits into topics and then can often see new subdivisions of the topics. (Perhaps marbles come in several shades of green. A child can develop subcategories, separating them according to their lime or turquoise qualities.) Thus the researcher can create subdivisions within the layers of categories; these can be conceptualized as a treelike structure.

But data are *not* like marbles! Your data documents are multifaceted, and you have precious knowledge about them. You are also likely to want to store information about people, places, sites, and so forth to do *descriptive coding*. Moreover, if your data are at all interesting, any passage will involve several, even many, topics, so topic coding is not merely a task of sorting into discrete heaps. Working on paper, you will need to copy a passage as many times as there are categories you wish to code it at, and finding patterns in that coding then becomes a challenge.

The use of computers for such research began with this challenge (L. Richards & T. J. Richards, 1994), and software has made it much more manageable. Working in software, you will learn quick ways of doing descriptive coding, storing, and using information. You will also find that once you start topic coding, it will be rare for you to code a passage only once. Specialized software makes it easy to code data as often as their meanings require and removes clerical delays in doing so. (For a detailed account of these processes, see Richards, 2005, chap. 5.)

Whatever your method, you are also likely to want to go beyond storing information and gathering material by topics. Coding always moves you to analysis by a process of creating and developing abstractions from the data. Once this is done, the category, rather than the data gathered at it, is the focus of attention.

Researchers often seek more abstract ideas or general themes in data. By a *theme*, we mean a common thread that runs through the data. Just as a theme melody in an opera emerges, recurring at different points, themes in data may keep "emerging," although their forms may not always be identical. You may identify a theme through processes of coding, or by stepping back from the data and asking yourself, "What is this all about?" In the latter case, you then return to the data and code portions that are relevant to the emerging theme.

The many functions of coding, then, link data with information, topics, concepts, and themes. These help to focus and conceptualize data as well as organize data so that they are malleable, allowing you to manipulate them as ideas and categories develop. Coding involves many processes, not merely "tagging" data with labels. Below, we address how you can get your data (and yourself) ready for coding and how you can store the ideas generated before and during coding in memos (and how you can manage memos so they work for you). We then describe several different ways of coding. In Chapter 7, we discuss the overall goal of abstraction, returning to memoing and annotating.

〰 GETTING INSIDE THE DATA

Coding for any purpose requires that you are familiar with the data and ensures that you get closer to the data. If the person doing the analysis has not conducted the interviews or done the observations, this task is far more difficult and more important.

In Chapter 5, we emphasized the importance of detail in recording data. When reading a data record or a portion of text, the researcher should be able to recall the setting and hear the participant's voice and inflection while reflecting on the meanings and implications of the text. Researchers who write their own field notes and transcribe their own data find "getting inside" the data much easier than do those researchers who have someone else transcribe the data.

Now, as you read a data record, read it purposively. You should be thinking not of that research event alone but of its relevance to your project. As you read, depending on your method and purpose, you will want to do many things. One is to record your ideas in annotations and memos. Another is to create and record categories representing what the text is about.

░ STORING IDEAS

The recording of ideas occurs in all methods. The process of *writing memos* enables you to reflect on the data record, or on the topic or theme, so your data management system should include an easy way of making and editing memos, which should then become part of the data set. Once you "know" the data, these ideas will come quickly.

In Chapter 7, we discuss the task of recording memos in more depth. As you start coding, you will often find that you wish to write about the new discoveries you are making in the data or the ideas you are developing from the data. Just do it! There are no rules about what a memo should be, or what is not a memo. New researchers often falter because they feel memos should be impressive or professional; experienced researchers use memo writing freely, to record their hunches and to think aloud. There are few examples of such memos in the literature because research reports normally do not cite the memos that contributed to the final product and the processes of writing up the research. Memos may be used in several different ways in order to help you get "up" from the data (Richards, 2005, chap. 4):

░ You may use them to record descriptions of events observed, the physical setting, or your memories of the mood or the context of a meeting.

░ You may use them to record your ideas or impressions about portions of an interview, and to link the text with the literature, with other data, or with very raw ideas that you do not want to forget— reminders of things to check out or to watch for in the future.

〰 You may use them to reflect on one word or phrase, on your annotations on discourse, or on your ideas about an entire document.

〰 You may use them to record your ideas about an idea, a category, or a theme at which you are coding data, or a concept you wish to develop.

〰 DOING CODING

Now start coding. Codes, at their simplest, are just labels. Coding data descriptively is like labeling jam or preserves; it will help you later to find whatever you are searching for (plum jam or any other variety). If you need somewhere to start, start here. But always ensure that you move beyond mere labeling.

As soon as you think about the labels you assign as a system, you are thinking analytically. If codes are organized in an index system, the labels work like cataloging in a library, allowing the user to find grouped-together material about related topics and showing how they are related.

But the labels on jars of jam convey no information about taste, and library cards cannot tell you whether or not a book is exciting. If you start with simple labels, start also thinking about the codes. During coding, data *make* the categories, in the sense that they alert the researcher to certain patterns and surprise with new meanings.

Coding is *linking* rather than merely labeling. It leads you from the data to the idea, and from the idea to all the data pertaining to that idea. For most, but not all, methods, it is important that the links lead both ways. Coding takes you away from the data—"up" from the data to more abstract ideas or categories. Coding will also take you "down" from the idea to all the material you have linked it to, and down from any of those segments to the whole document. Coding, if well done, is the way you monitor occurrences of data about your ideas and the way you test them. It makes resilient links between data and ideas, links that you can trace back to find where particular ideas came from and what data are coded there, to justify and account for the interpretation of the ideas.

Coding is also a way of *fracturing* data, breaking data up, and disaggregating records. Once coded, the data look different, as they are seen and heard through the category rather than the research event. This is both a great advantage and a danger. The retrieval from a code offers a new focus, the ability to compare and be surprised by things not seen when the data documents were viewed as a whole. But it also wrenches the data segments out of context, distancing you from the original whole.

Descriptive Coding

What Is It Used For?

Descriptive coding is used to store things *known about* data items (e.g., respondents, events, or contexts). The researcher can then access this factual knowledge about the respondent (gender, age, and so on), the setting (in the hospital, clinic, at home), or context (year of interview, which question was being answered, and the like) when seeking patterns, explanations, and theories.

Descriptive codes "entail little interpretation. Rather, you are attributing a class of phenomena to a segment of text. The same segment could, of course, be handled more interpretively" (Miles & Huberman, 1994, p. 57). Descriptive codes are used rarely for simple retrieval, but often for asking questions of the data. (Did women and men see an event differently? Did younger women have a different concept of attractiveness than that of older women?)

How Is It Done?

Ideally, descriptive codes are incorporated into the data management process, with each new respondent or site allocated all the characteristics that are relevant. Working manually, you will need a fact sheet or ID card for each. Working in a word processor, you might put this information in a header or footer. If you are using software designed for qualitative research, you will code data documents with their attributes or, even easier, import the information automatically from a table.

But be forewarned: You should store as much information as you need, but no more. Overcoding confuses. Coding should be sufficient to ask and answer your question. If you are using a computer, you can easily import further information later if you need it.

Where Is It Used?

Descriptive coding is common, if not universal, in all methods, given that qualitative research requires awareness of context. But methods differ in their emphasis on descriptive coding. Researchers using theory construction methods tend to do little such coding, because the details about individual cases will be subsumed in the development of an overall theory.

In the earliest stages of research design, you should consider what information you need to store (about people, sites, settings, and so on). What do you want to be able to ask? And if you are intending to work with mixed methods, what data do you want to move between qualitative and statistical analysis?

Topic Coding

Topic coding is the most common and the most challenging sort of coding done in qualitative research, especially since computers make it deceptively easy. In almost any project, the researcher will, at some stage, need to be able to access data by topic. However, you should never allow qualitative topic coding to become what Lyn Richards terms *data disposal* ("This is about this, this is about that . . ."). Topic coding is a very analytic activity; it entails creating a category or recognizing one from earlier, reflecting on where it belongs among your growing ideas, and reflecting on the data you are referring to and on how they fit with the other data coded there.

What Is It Used For?

Topic coding is used to identify all material on a topic for later retrieval and description, categorization, or reflection. This sort of coding can be fairly descriptive (the respondent is talking about the headmaster) or more obviously interpretive (hostility, authority figure, role model, and so on).

The purpose of topic coding is seldom merely to allow the researcher to find material according to label. Sometimes its purpose is to provide accurate descriptions of the varieties of retrieved material, and sometimes it is to provide new ways of access to data, by combining codes in sophisticated searches (What did the women say about the headmaster if they were also hostile to traditional formal authority systems?). Think of topic coding as coding *up* from the data. It easily becomes analytic, because you can review data coded at a topic for dimensions or patterns, coding *on* from there to new, finer categories.

How Is It Done?

You can achieve topic coding by marking up printed text manually (using colored lines and topics in the margins, or by cutting and pasting), but this method is satisfactory only for very small projects. It is slow, and it does not result in material that is easily accessible for you to review and recode. Using index cards or folders for each topic works better, and for some this method provides a useful way to learn how to code. But such coding processes sever data from the context and impede your thinking further about the topic (Did the young women talk differently about class than older women?). Even such a simple question requires difficult re-sorting of data extracts. Topic coding using manual methods will fracture data by removing it from sequence. This may work against some method-ological goals, such as identifying processes in grounded theory.

Using a computer allows you to work directly with the text, selecting passages and doing coding onscreen. You can then immediately view what is coded there "live" or view the context if you wish to return to the original document. You can work on the material gathered by topic to create new categories and to make finer distinctions in the process of "coding on." Or, of course, if the data to be coded can be mechanically identified (e.g., if they consist of any particular series of words), you can use the computer to code words and context automatically.

Where Is It Used?

Topic coding is used in almost all qualitative research methods. It is nec-essary in any project where there is emphasis on finding *all* the data about an aspect of the site or experience studied, or on accurately portraying the distribution of different attitudes, experiences, and so forth. Ethnographers piece together accounts of social processes. Researchers working in both discourse analysis and grounded theory code data using words that occur in the data. In grounded theory, these are termed *in vivo codes.* Narrative analysis requires that the story be seen whole but also that the researcher be able to extract topics. (Thus you might wish to topic code to gather all the material, for example, on class in a series of life stories.)

All methods benefit from broad topic coding early in a project (What, if anything, do we have on this?). Topic coding may also be useful as a first stage of the analysis, when the researcher is exploring to see "what's here." You may use more finely tuned coding once categories are firmed up and you understand "what is going on." And almost all methods use topic coding as a first step to more interpretive coding. Reviewing all the material on a topic, you may see subtler subtopics or dimensions (e.g., Does *hard life* have different meanings in these narratives?). Create new categories for these on the way to analysis.

Analytic Coding

As you begin to code for more categories, topic coding becomes more analytic. Perhaps you were coding every time "a hard life" was mentioned in the narratives of elderly people looking back over their histories. The stories alert you to the significance of "hard" in this context, and you want to store the insight that "hard" and "easy" may shape memories. Make a special place or a special way of seeing the categories that are growing in complexity, and start writing your queries as memos. Later, if something happens to make you wonder if "having it easy" is always seen as a benefit, you have not preempted this discovery by making it "belong" with "looking back." But what you do with this discovery will of course depend on what you are asking and the method you are using.

What Is It Used For?

Analytic coding is used to make, celebrate, illustrate, and develop categories theoretically. It is labeled *analytic* because in creating categories you go *on,* not just linking them to the data but also questioning the data about the new ideas developing in the new codes. The purposes of analytic coding include the following:

⧈ To alert you to new messages or themes

⧈ To allow you to explore and develop new categories or concepts

⧈ To allow you to pursue comparisons

Usually, the memos you write will contain references to many instances of recurrent themes. The links that connect these to the original sources of the ideas (if they were ever there) may now become unimportant. Think of analytic coding as *taking off from the data.*

How Is It Done?

Analytic coding helps you to develop themes or categories. You may make new, more general categories or simply start memos about possible categories. But however you store the ideas, be sure to write about the data and rewrite, linking more data as categories recur or as you see them differently in the process of abstracting. The categories grow in your memos. Date your entries within particular memos, or hyperlinks if you are on computer; keep a log trail (Richards, 2005, pp. 22–23). Summarize and synthesize data as the categories develop.

Read accounts written by people who have done analytic coding (such as Dey, 1995; Miles & Huberman, 1994; and the highly original account of coding in grounded theory method by Turner, 1981). Review your codes for ones that may potentially contribute to a higher level of analysis. Seek to generate new categories and read how others have done so. Validate your development of a conceptual scheme with new data.

Where Is It Used?

Most qualitative researchers who are seeking to develop theory will do analytic coding. But as methods differ, so too will the processes. If in doubt, aim at working *up* from topic coding, then working *out,* by coding around the topic to establish its significance and meaning.

When coding is used for category construction, the emphasis is less on the labeling of text and more on the evolving categories. Researchers may (but often do not) identify segments of text as belonging to codes, but what they emphasize is the ability to discover and develop categories from data.

This is territory that grounded theory method dominates, and, sadly, it is now much-disputed territory. Coding is central to the arguments about grounded theory that we mentioned in Chapter 3. Texts from Strauss's "school" offer techniques of a *coding paradigm.* These provide the researcher with ways of interrogating categories produced in response to data, asking how they link to other things the researcher knows. The terms used suggest the emphases: *open coding* is coding aimed at opening up the data, identifying concepts that seem to fit the data; *axial coding* moves the focus around a concept; and *selective coding* offers intense analysis that focuses on one category at a time. This sense of hunting down central themes is very different from the emphasis on the less structured processes of "theoretical sensitivity" in Glaser's writings, and a central thrust of Glaser's (1992) attack on Strauss's work with Corbin (1990) is that their methods involve "forcing data."

If you go on to use the grounded theory method, you will find these issues hotly debated, and you will need to deal with them. The central goals of making concepts and theorizing about concepts are, however, common to both sides. Both Glaser and Strauss emphasized that coding should start early, a message central to their early work together. Their works advise researchers to code by building cumulatively on coding and being strongly aware of the internal development of categories and "changing relations between the categories" (Glaser & Strauss, 1967, p. 114).

Our advice is that you seek an understanding of the purposiveness of open coding and get a sense of what it can do to data. This is splendidly portrayed in Strauss's (1987, chap. 2) book *Qualitative Analysis for Social Scientists*, which contains transcripts of tapes from team sessions during which researchers worked with data. As so often occurs in qualitative research, the actual performance proves much less rule bound than textbook instruction would lead you to believe, and we are reminded yet again that apprenticeship is the perfect way to learn these skills. (For an adaptation of some techniques of open coding, see Richards, 2005, chap. 5.)

〰 THEME-ING

Coding to develop themes may occur during any of the processes discussed above, but by *theme*, researchers usually mean something more pervasive than a topic or category. As we have noted previously, a theme runs right through data and is not necessarily confined to specific segments of text. However, once a theme is identified, you are more likely to see segments of text that are pertinent to it.

Discovery and coding for themes usually involves copious and detailed memos that are abstract and reflective. These memos must be categorized and sorted, as they, too, are data. As you code supporting data, a coterie of codes may develop around the themes. We will return to theme-ing in Chapter 7.

〰 PURPOJIVENEJJ OF CODING

The purposes of coding are very different for different methods as well as for different stages in a project, but all coding should have purpose. Before you code, ask, "Why am I doing this?" This is a particularly urgent message to those using computers to aid in their research. Because the computer makes coding simple, researchers who are unclear about what they should do next are tempted to continue coding (and perhaps making data) rather than move their studies forward conceptually. Coding should never become a routine process of data disposal (considered as something to "get through")—and if you find that it does become routine and mechanical, do something else. Then, after your interest in working has returned, ask yourself why coding had become boring.

Note that not all coding is aimed at retrieval of all the material coded. Indeed, of the types of coding described above, only topic coding normally leads to retrieval of all the text coded at a category, and such retrieval is then usually a means to an end, a way toward pattern finding, exploration and reflection, and coding on to create new categories. Being able to get back everything coded within a category is rarely an end in itself. Sometimes you can achieve the research goal more simply by comparing two or more codes to locate patterns. For instance, a researcher conducting a focus group analysis to identify areas of concern might create a new code for each topic of concern raised in the group and code it at each segment of the transcript referring to that topic. The researcher's role may be limited to reporting that this is a topic of concern, citing the participants' own words. But even in a very limited study, they will normally aim to say something more useful (e.g., about the range of attitudes on the topic or the ways these attitudes are patterned by individuals' characteristics, such as gender or age group).

Coding may not be primarily for retrieval but for access, as in discourse analysis, where location of segments by keyword is a first step toward gathering all the material where one of a set of related terms occurs. Thus, in a study of a television campaign, Potter and Wetherell (1994) used computer text search to find occurrences of any phrases that referred to cure rates. Retrieval of all such passages together allowed them to reflect further on the discourse used in the campaign. They comment, "If we were not sure if the sequence was relevant, we copied it anyway, for, unlike the sorts of coding that take place in traditional content analysis, the coding is not the analysis itself but a preliminary to make the task of analysis manageable" (p. 52).

☷ TIPS AND TRAPS: HANDLING CODES AND CODING

Ways of managing, reviewing, and processing data are typically the least-noticed stage of research and the least written about. Many texts, and indeed many studies, hardly mention these techniques. Instead, they emphasize ways of making data and the end results—the theories derived from data. Because of this silence on the subject, researchers are offered little instruction about techniques of coding or the different ways of coding that are appropriate for different purposes. Although many authors note that coding can create an enormous clerical load, few add to this warning that because computers code so easily, using computers can put qualitative researchers at risk of what Richards (1995; T. J. Richards & L. Richards, 1994) has called "coding fetishism." Coding is central to analysis, and analysis should never be routine and mundane. So how do you deal with these issues?

Code as You Learn

In all the different methods, coding should be started early. As data come in and you think about them, prepare them for analysis and code them. Avoid letting data build up and then coding in a block. A huge backlog of data makes them seem inaccessible, and should you lose the pacing of data and analysis, it will be harder for you to see thin areas in your data or to use what you find to guide future data making.

Always See Coding as Reflection

Don't allow yourself to see coding as a stage of data preparation *prior to* thinking—coding is a theorizing activity. Even apparently simple topics (such as popularity) may sprout complex ideas. Keep revisiting each category and reviewing what is coded there, and whenever it is surprising or interesting, write a memo.

Never Code More Than You Need

How do you know how much is too much? It helps if you don't see the coding of a document as your last chance to read and think about it. Your data management system should allow you to return easily to the context or the full document—to think and maybe code some more. Not many things in qualitative research are done only once, at one particular stage of the project. Given that you can always revisit what you've done, our best advice concerning descriptive and topic coding is that you code for anything you are likely to want to ask questions about.

For analytic coding, our advice is different: If it moves, code it. Storing and revisiting these categories is likely to be what makes your project come alive. With skilled use of your computer tools, you can flit from the code to the memo about it, to the material coded there, to the original document, and to the all-important theme, altering the coding and reviewing it, creating new memos, and coding them.

Manage Your Codes

One commonality among all forms of qualitative research is that analysis does not stop with coding. This is why the early computer programs, when seen basically as coding machines, were widely rejected. "Codes *are* theoretical directives," Strauss commented in a meeting with Lyn Richards in 1995. "Codes are the crosscuts between talk and biographies. . . . Coding is putting interpretative structure on the data." There is no point in making categories for their own sake—they are for linking to data and to each other.

This returns us to the issue of data management. Good data management is essential in any project in which the number of codes is too great for the researcher to be able to find any given code instantly. To use codes inconsistently, or to shift the interpretation of them, is to invite disaster. Think of coding categories not as receptacles for data but as concepts you work with. If you are topic coding, get in the habit of visiting the category when you code into it. Are you being consistent? Are new themes turning up? How much variety is there within the category? Should the category be developed into two finer categories? Do its definition and its memos need revisiting? If you are working in a team, you will want to check from time to time that you and your colleagues are using each category the same way, and discuss it if you are not. Coding is a process of category refinement, even if the coding is solely descriptive. Remembering this will help you to avoid treating coding as a substitute for thinking. It will also protect you against the perils of excessive coding and the boredom of unnecessary clerical coding.

When you reflect on a category, locate it in some way that helps you find it again. If you are working manually, you may do this by putting the category file in a logical place in a filing cabinet. If you are using the computer, when you discover categories that seem to go together, use your software's ability to make "trees" of categories and subcategories so that you can see them and manage them efficiently (Richards, 2005, chap. 6).

Monitor Coding Consistency

Because qualitative coding is primarily interpretive, two researchers will rarely produce identical coding. Why should they? Coding reliability tests can establish reliability of descriptive and simple topic coding (What is her age group? Have I coded every time she has mentioned her treatment?). In quantitative inquiry, researchers seek to have far more confidence in consistency; they want to ensure that they have the right numbers down and have no omissions. But differences in interpretive coding indicate only differences in researchers' purposes and perceptions (Richards, 2005, p. 98).

If you were to observe a quantitative research team at work, you would find that arguments occur (Is this a 3 or a 2?), but inconsistencies are obscured because there is no record, for example, of the wording of the statement interpreted as a 3. Good quantitative researchers, aware of this problem, work very hard to clear up inconsistencies in interpretation and to record uncertainties. Good qualitative researchers handle inconsistency in a different but related way. Different interpretations are inevitable, so you must monitor, revisit, and debate them, and make them a part of the process of analysis. This process, in itself, can lead you to important insights as you get beneath taken-for-granted meanings.

Under two conditions, however, it is essential that you check the reliability of your coding:

〰 You should build into your research design ways of checking your consistency of coding over time and therefore your understanding. Document and track your use of these methods.

〰 If you are working in a team, you need to build in a process for monitoring the consistency of coding by different team members. We would not expect identical coding by different people, and if we found it, we would worry, because it probably means that nothing interesting or ambivalent is being coded. But it is essential that you know the different ways in which team members interpret categories and handle data, and that you monitor and discuss the differences.

〰 UJING YOUR JOFTWARE FOR CODING

In Chapter 4, we introduced the ways of gathering the preliminary material that lead to research topic and question, and using the computer to handle and manage data records as data are made. Now we go on to coding. The computer can take much of the clerical burden from each of the modes of coding—but it leaves you with the task of interpretation.

Approaches

All software packages support coding, but they support it in different ways. Many have a choice of ways of coding. In Chapter 5, we discussed the coding discussed in this chapter as "descriptive coding." This is usually done by import of attributes and their values for the relevant documents or cases. Here, we come to the aspects of qualitative research, which from earliest software development have been the focus of computer assistance—topic and analytical coding.

Software will support the following processes, which you need to learn:

≫ How to do topic coding by making and defining categories and managing them as they develop

≫ How to do topic coding automatically, by section or by text search

≫ How to do analytic coding, which includes discovering and developing categories, linking them to memos, and using the computer to support the discovery and exploration of themes

≫ How to review the data coded at these categories, recode them, and develop them

≫ If you wish to try doing it in software, go to Appendix 1, where there is a guide to tutorials available online.

Advances

You don't need a computer to code. Before software, coding was done with pens and filing cabinets. The central process is interpretation, and computers don't interpret data—researchers do.

So what is different about coding if you do it with software? For the researcher, there are three critically important differences:

≫ Computers code very easily and swiftly, so coding with software is much faster and more efficient than on paper. Your interpretation can be far more easily and immediately stored as coded data.

⧹ Computers can store far more information than paper files. There is effectively no limit in good software to the number of coding categories you can make or the amount of data you can code at them.

⧹ Coding data—that is, the categories you create and the selections you wish to code at them—is stored by software as pointers to the coded segments, not as marked-up or cut-up extracts of text. This information is very much more flexible, and much more easily altered, than paper records.

⧹ Unlike the filing cabinet, software can very easily take you from the coded segment to the context. Some software can show you all material coded at a topic or concept, to allow you to rethink, revise coding, and code on to further, new dimensions of the concept.

⧹ With software tools, you can ask questions about patterns of coding that were literally impossible with paper records. Qualitative researchers usually need to go beyond their first coding. It's not enough to get in one place everything said about a problem. You are more likely to want to know, for example, when people were coded as saying they had this problem, what did they say—anywhere in their interview—about their trust of this advice?

Alerts

The greatest dangers of qualitative computing lie in just the facility, capacity, and patience of the computer! Coding is essential to most methods, but it can become a trap if you are not aware of these risks.

1. Never let coding become just a clerical duty; if it's qualitative coding, it has a purpose. Why did you create this category? What are you planning to do with it? Coding is not a substitute for interpretation but an expression of it.

2. Keep coding in its place. It should not dominate any part of your research timetable. When you are unsure what to do next, or when the meaning is not "emerging" from your data, it is very easy to code some more.

3. Don't overcode. Take care to avoid the dangers of "coding fetishism" (Richards, 2005, pp. 100–101), compulsive activity that happens when researchers feel they can't think about data unless it is coded. Coding compulsively can easily replace reflection and exploration of data.

4. Don't allow coding to stop you from doing other interpretive processes. Whenever you are coding, aim to store changing ideas at the same time, and to reach for other tools to write memos, revisit coded data, and ask about what your coding is uncovering.

◊ SUMMARY

Coding is the strategy that moves data from diffuse and messy texts to organized ideas about what is going on. According to the researcher's analytic goal, coding may take one of several forms: descriptive coding to identify information according to topics, analytic coding to facilitate interpretation, or coding to identify themes. A researcher may use one or more types of coding simultaneously.

Coding enables data retrieval so that you may begin processes of analysis, but analytic coding also enables you to ask questions of the data. Coding is a cognitive activity, not an automatic function—be certain to think as you code. Do not overcode, and be aware that analysis begins with coding—it is not the only component of analysis. In Chapter 7, we discuss the thinking processes that are the reason for coding and the results of it—categorization, conceptualization, and abstraction.

◊ RESOURCES

Richards, L. (2005). *Handling qualitative data: A practical guide*. London: Sage.
Chapter 5 gives detailed advice on doing coding and using and checking computer coding.

Other Resources

Bazeley, P. (1999). The bricoleur with a computer, piecing together qualitative and quantitative data. *Qualitative Health Research, 9,* 279–287.
Boyatzis, R. E. (1998). *Transforming qualitative information: Thematic analysis and code development*. Thousand Oaks, CA: Sage.
Coffey, A., & Atkinson, P. (1996). *Making sense of qualitative data*. Thousand Oaks, CA: Sage.

Coffey, A., Holbrook, B., & Atkinson, P. (1996). Qualitative data analysis: Technologies and representations. *Sociological Research Online, 1*(1). Retrieved May 8, 2001, from http://www.socresonline.org.uk/socresonline/1/1/4.html

Dey, I. (1995). *Qualitative data analysis: A user-friendly guide for social scientists.* London: Routledge.

Glaser, B. G. (1978). *Theoretical sensitivity: Advances in the methodology of grounded theory.* Mill Valley, CA: Sociology Press.

Glaser, B. G. (1992). *Basics of grounded theory analysis: Emergence vs. forcing.* Mill Valley, CA: Sociology Press.

Potter, J., & Wetherell, M. (1994). Analyzing discourse. In A. Bryman & R. G. Burgess (Eds.), *Analyzing qualitative data* (pp. 47–67). London: Routledge.

Richards, L. (2005). *Handling qualitative data: A practical guide.* London: Sage.

Richards, T. J., & Richards, L. (1994). Using computers in qualitative research. In N. K. Denzin & Y. S. Lincoln (Eds.), *Handbook of qualitative research* (pp. 445–462). Thousand Oaks, CA: Sage.

Strauss, A. L. (1987). *Qualitative analysis for social scientists.* Cambridge, UK: Cambridge University Press.

Strauss, A. L., & Corbin, J. (1990). *Basics of qualitative research: Grounded theory procedures and techniques.* Newbury Park, CA: Sage.

Tesch, R. (1990). *Qualitative research: Analysis types and software tools.* London: Falmer.

Wolcott, H. F. (1994). *Transforming qualitative data: Description, analysis, and interpretation.* Thousand Oaks, CA: Sage.

7

Abstracting

One of the few commonalities among qualitative methods is that all methods urge the researcher to start analyzing as soon as the research begins. There are good reasons for analyzing data concurrently, as soon as it is made. In earlier chapters, we stressed that all qualitative methods aim to build understanding from the data and that research design adapts to growing understanding. This means that abstracting begins at the beginning of a project. Completing the gathering of all the data and *then* thinking about them is usually highly problematic, because this does not allow the data-gathering process to be data driven. Doing so, however, is also very attractive if you are unsure of what you are doing. It's much easier to do 10 more interviews than to confront the challenge of finding out what is to be learned from those already done. This means that as you start making data, you need a strong idea of how you will analyze it.

All research shares the goal of making the researcher think abstractly. *Abstract thinking* involves transforming data from individual instances by creating, exploring, and using general categories that are derived from the data. In qualitative research, this is normally a primary goal, even a defining condition for a project to be considered qualitative. As we have shown in our sketches of methods in Chapter 3, all of the qualitative methods we are discussing in this book seek a sense of understanding, of things coming together. It is this process that distinguishes description from analysis.

Abstracting from data gets you somewhere else—away from the data and toward the concepts that help you understand them and (sometimes) build theories about them. Consider the qualitative studies you have found satisfying or intriguing—what was satisfying about them was probably an abstraction. Rather than summarizing the experience of all the patients in a cancer ward, Glaser and Strauss (1971) developed the idea of "status passage." After talking to many people in different kinship situations, Finch (1989) wrote not about their individual experiences but about family obligation. Sometimes abstraction is driven by concepts generated from the data. Morse developed the concept of *compathy* to describe the sharing of another's pain experience (Morse, Mitcham, & van der Steen, 1998) when her interviews with nurses who worked with major trauma patients described feelings for patients that did not fit the description of empathy in the literature. Sometimes such abstraction is informed by theory about the phenomenon studied. Richards (1990), in her analysis of neighbor networks, drew on the feminist literature on women's work and ideology of family to interpret conflict in local associations as labor on the border of "private and public" worlds.

How abstracting is done is far too often presented as a mystery. A constant theme in the literature is that the theory emerges from the data, but we have never been privileged to see such an apparition. Whether driven from the data or from other theory, qualitative abstraction always requires the researcher's active exploration of data. Understanding is developed as a result of the researcher's insight and the researcher's work. As we have emphasized earlier, the research process is not passive. The researcher drives it, making informed decisions, thinking, linking, and abstracting. If theory emerges, it is because the researcher "emerged" it.

Understanding the researcher's agency is a first step toward being able to do it. You don't need supernatural powers or the touch of a guru to be able to think abstractly about your data. You do, however, need good research design, good data, skill, and creative and concentrated efforts at thinking. We cannot stress strongly enough that an adequate account of the data or answer to the research question will emerge only if the right conditions exist. For any method, abstraction requires that the data be in the right state, that the researcher be receptive to the emerging ideas, and that the researcher understand and undertake appropriate analytic strategies.

Techniques of abstracting from data vary widely. The terms associated with abstraction include *classifying, coding, distilling,* and *seeking "themes."* To learn specific techniques, go to the specialist literature. Here we focus on what the various techniques have in common: the first steps of abstracting "up" from the data, which are usually called *categorizing* and *conceptualizing.* All qualitative researchers aim to create categories that are more general, drawing together the complex immediate messages of the data in more abstract topics or groups, and most aim to move from this sorting of data to more theoretical concepts.

☰ THE FIRST STEP: CATEGORIZING

Categorizing is the first, minimal step to abstracting, but it is often also an end goal. For some projects, it is a first step toward creating theory. For others, the goal is *only* categorizing and category exploration, and the analysis process in such a study may require nothing more abstract than giving a very accurate account of what is going on. The ability to offer "thick description," or to locate a surprising pattern, may transform complicated data into a story that makes sense—but doing that requires categorization.

It is important to understand that the research question sets the goals for the outcome of the project. In this area, there is a serious lack of fit between what is actually done in the research world and the literature on methods. Either implicitly or explicitly, many texts overlook or devalue studies that do not progress from categorizing to theory construction. This is a major source of confusion for novice researchers, who find it hard to recognize in the texts the sorts of studies they are seeing presented at conferences. As we mentioned in Chapter 3, many such studies seek patterns rather than theories. Such a project does not conform to one of the recognized methods of theorizing analysis, yet researchers are led to believe it must still be labeled with one. If your goal is simply to find out if anything new came from open-ended questions in a survey, or to evaluate a pilot focus group, there is no need to pretend that it is grounded theory or phenomenology. But there is a need for abstracting.

Categorization and Coding

All qualitative research involves categorizing (as does almost all reflective thinking). Only some requires coding—the set of processes described in Chapter 6—for gathering the material *about* the categories. Thus categorizing only sometimes leads to coding.

Coding, on the other hand, *always* requires categorizing. As we have shown in Chapter 6, in any of the ways of coding, the researcher needs to say, "This is a topic or idea" on which he or she will gather material. Coding generates categories. During coding, the data "make" categories as the researcher is alerted to concepts, themes, patterns, and surprises with new meanings. During coding, new categories seem to "happen" when an existing code doesn't quite fit the data or new data suggest several dimensions of the category. All qualitative methods that involve coding treat it as category creating and at least one way of abstracting *from* data (Richards, 2005).

Coding also supports development of categories. If paper records are coded, they can be gathered in heaps, reread, and reviewed. If computer software allows "live" display of coded material, the researcher can rework a category, redo coding, and create related categories as he or she explores the material.

But many qualitative researchers think about the data without coding. As we showed in Chapter 6, "coding" has as many purposes and as many recommended techniques as there are methods. We have warned, then, against coding in the absence of methodological purpose.

Categorization as Everyday Strategy

When our goal is comprehending and learning, seeing or saying something new, predicting or understanding, we categorize. So for most humans, and for some qualitative researchers, category construction is a normal process of comprehending that has nothing to do with coding. Only sometimes is it important to gather material about a category.

Qualitative thinking is very much like everyday thinking. Researchers categorize constantly, creating and using general ideas about particular data items. If you have trouble with the task of category creation in your research project, spend a day watching your processes of understanding the complex data of everyday life. We manage these processes largely by categorizing. If somebody approaches you on the street, you will sometimes move to greet them or offer assistance, or sometimes do your best to avoid contact. Between the approach and the response lie categories—what sort of person is this, what sort of approach, and of what sort is this attitude? Some answers are so obvious that we take them for granted. For example, this person is a man (a nonfemale human of mature age). There is no subtle interpretation here (unless the person is heavily disguised or obscured), but we have categorized him. Male gender is a category that allows us to think of all males and ask questions about male behavior or male experience. Whether someone is a "homeless man" or a "businessman" will involve more interpretation: How, from physical appearance, do we "know" this person is homeless? In addition, "homelessness" is a category to which people attach stereotypes that may have nothing to do with experience or fact.

Categorizing is how we understand and come to terms with the complexity of data in everyday life. It is our normal making-sense activity, and it is efficient. If one thing is a "sort of" something else, we can respond to *types* of experience, to patterns, to predictable behavior. Categorizing is also always risky, because it obscures uniqueness, preempts discovery, and may blind us to the unusual. (Oops, that "homeless man" is an eccentric millionaire!)

It is odd, then, that a process so normal in life and in research can become a major obstruction to doing qualitative research. But category construction can, in our experience, be the earliest obstacle, and sometimes a fatal one, for novice researchers. Understanding (properly) that categories are the first step to theories and having learned to treat theories as the products of Great Minds, new researchers can easily suffer failure of nerve when they think they see something in the data but it seems too trivial to record. Our advice? Record it!

⬥ THE NEXT STEP: CONCEPTUALIZING

In qualitative research, categorizing is how we get "up" from the diversity of data to the shapes of the data, the sorts of things represented. *Concepts* are how we get up to more general, higher-level, and more abstract constructs. Concepts are mental images. Researchers usually seek ways of moving from categories to concepts and then of building frameworks of concepts that map or image the subjects of research. As Miles and Huberman (1994) note, "Theory-building relies on a few general constructs that subsume a mountain of particulars."

Categories may never develop into concepts, but to the extent that they do, you move the study from description to analysis. As you come to understand and go beyond complex data, how far you go beyond is a question decided by your method. Even the simplest categorizing is analytic. The processes of making a category involve discovering a new idea and naming it, storing thoughts about it, managing its relation to other categories, holding it in mind, and linking it in the growing understanding of your work—all of these are theorizing processes. They are the first uncertain steps by which the researcher sets out toward the hazy goal of theory. With categorization, the researcher puts a first foot on what Carney (1990) calls the "ladder of abstraction."

⫸ DOING ABJTRACTION

Qualitative methods of all varieties share a commitment to allowing categories to emerge from the data. For many researchers, this is almost a defining characteristic of qualitative method, and it is certainly one of the few commonalities among methods. (If you were not trying to learn from the data, you probably would not be working qualitatively.) In most methods, some categories are decided in advance, but all methods seek new categories, which emerge as the researcher achieves understanding of the rich data.

There is such a variety in the sorts of categories you are likely to create in a qualitative study, in how and when they arrive in that study, and in the uses to which they are being put that it is at first easy to see them as unrelated tools for different purposes. In the following subsections, we outline the range of ways in which qualitative researchers make and use abstract categories for describing and analyzing data. Not surprisingly, this relates to the range of coding techniques.

The fit of a particular method with a particular research question is often best observed when the researcher considers the ways of abstracting. Methods differ very clearly on each of four dimensions that are very relevant to the location of the project and the preparation of the researcher. Each method has its own answers to these questions: When does abstraction happen? Where do abstractions come from? How are abstractions created? What analytic outcome is being sought? Like the other tables in this book offering comparisons among methods, Table 7.1 is a sketch map; it is intended to encourage you to undertake more detailed reading on your selected method and to help you put your method in context. We suggest that you read it in conjunction with the methodological map provided in Chapter 3.

When Does It Happen?

At least some abstraction will always happen early. One of the myths we have challenged is that a good researcher always approaches data with an open mind, and that this means it should be a mind empty of knowledge about the topic. This myth can seriously hinder the researcher who accepts it, and then finds it untrue of his or her situation. If you deny that you know a lot about the situation you are studying (and many

Table 7.1   Doing Abstraction in Three Different Methods

Method	When Does Abstraction Occur?	Where Does Abstraction Come From?	How Is Abstraction Done?	What Is the Goal of Abstraction?
Phenomenology	Not until one has the data: Previous ideas and knowledge are bracketed.	Themes and meanings in accounts, texts	Deep immersion, focus, thorough reading	To describe the essence of a phenomenon
Ethnography	Prior knowledge of site, situation; understanding develops during field research.	Knowledge of social and economic setting; observation and learning from the setting	Rich description; combination of qualitative and quantitative patterning; coding, comparing, reviewing field notes	To identify themes and patterns; to explain and account for a social and cultural situation
Grounded theory	Abstraction is from the data, but can be informed by previously derived theories.	Categories derived from data (observations or line-by-line analysis of texts); constant comparison with other situations or settings	Theoretical sensitivity; seeking concepts and their dimensions; open coding, dimensionalizing, memo writing, diagramming	To identify a core category and theory grounded in the data

researchers do), you are likely to walk those assumptions into your study without acknowledging them.

Qualitative methods—and teachers of qualitative research—differ in their approaches to the handling of prior thinking. You may be advised or urged to assess your prior ideas and theory from the literature at an early stage in the project, as a means of shaping data making and designing access to data. If your method and context require this, the research design will involve reviewing and using this prior knowledge. Using computer software to categorize literature as well as project data will assist you in this task. Be aware, however, that some methods, for good reason, explicitly discourage researchers from categorizing before they explore their data documents. This is because these methods emphasize having the meanings that occur in the data determine the processes of abstracting from those data. This is the reason for the literature on "bracketing" of prior knowledge.

The handling of prior knowledge, then, is a clear example of the need to locate your project methodologically. Nobody approaches any social situation free of categories, and often alongside the categories for thinking about a particular situation come concepts from the literature and beginning theories. So you need to be aware of the pathway to analysis in your chosen method and of the timing of analysis processes. There are strong differences in the acceptability of prior categorization, but in no method is abstraction *entirely* a priori. None of these methods, for example, will look to answers to a "precoded" questionnaire for abstraction. (This term is in fact a misnomer, given that the survey researcher is not preemptively *coding* the interview to be conducted, but *precategorizing* the answers.)

Where Does It Come From?

This is the area of greatest commonality among the methods. As Table 7.1 shows, all three of the major methods anticipate that abstractions come to some extent "up" from the data rather than "down" from prior theory. This commonality can be a source of confusion. All these methods, for example, rely largely on text. Some methods vary in that they use videotapes or photographs, but they all also use text. There are certainly variations in practice—for example, regarding whether an interview is transcribed and, if so, how complete a transcript should be and whether hesitations, tone, and body language should also be recorded—but few would not keep a record of the interview at all. Thus all require data that can be rich enough to generate interpretation and support abstraction.

How Is It Done?

Researchers employing different methods have different terms for this qualitative process of category discovery. Watch for accounts of the researcher's agency. Researchers talk of "hearing" themes, "feeling through" meanings, and "seeing" patterns. These phrases hide practical skills—the things the researcher does to data.

When you approach a method, considering its relevance for your project, it is very helpful to locate practical, honest accounts of what other researchers have done to their data. These are often missing from published research reports. For examples of some practical accounts that include this information, see the list of resources in Chapter 3.

What Are You Aiming For?

All of the processes of abstraction involve work. It is useful to distinguish, following Turner (1981), between theory emergence and theory construction. Theory emergence is often an event of discovery; theory construction is a craft. The accounts of studies and stories of researchers emphasize the ongoing work of piecing together clues and themes, tiny bits of understanding, and recurrences and links. The qualitative researcher builds a "network of concepts, evidence, relations of concepts, coordinations of data, of hierarchies of grain size where the theory/data/explanation chunks of one grain size are the data for the work of the next grain size up" (T. J. Richards & L. Richards, 1994, pp. 448–449; see also Lofland & Lofland, 1995). For a more detailed discussion of possible outcomes of your project, and possible aims for your analysis, see Richards, 2005, Chapter 7.

ॐ MANAGING ABSTRACTION

The processes discussed so far make another layer of data. The first effect is that you now have not only a growing body of rich data but also a growing body of categories, concepts, and thoughts about them. Almost any qualitative project rapidly acquires a growing task of managing elements so they can be known, viewed, explored, and linked. Some of the literature will assist you in these processes (on manual methods, see especially Miles & Huberman, 1994; if you are using software, see Richards, 2005).

Whatever your method, we recommend early attention to ways of managing not just data documents but also ideas, which are your most precious data.

Documenting Ideas: Definitions, Memos, and Diaries

For novice researchers, often the greatest barrier to thinking up from data is the sense that category construction is a majestic and magical performance. The best way to climb over that barrier is to work actively with and within your categories. Make memos about them generously—these can always be deleted later. Even the simplest categories grow and change during a project, and you must record these changes if you are to be able to tell the story of your analysis later.

For any idea, topic, theme, concept, or other abstraction, it is advisable that you store somewhere a description of how it is to be seen and used. When it changes, that description can be changed, and if the change is significant, it can be tracked. As ideas about a category grow, it is useful, sometimes essential, that the category acquire a memo. As your confidence grows, so does the memo, possibly accompanied by or linked to other memos.

Growing Ideas

A theme of this book has been that ideas emerging in a project grow with the data. Whatever your method, it is important that you expect and encourage your ideas to change and be ready to record changes. This means that idea management must be flexible and introspective. As you continue to work with data, you will need ways of revisiting and rethinking categories to develop them further and refine their dimensions (see especially the techniques for "dimensionalizing" in grounded theory, Strauss, 1987, p. 180). You also need ways of challenging and assessing the themes you are seeing in the data.

> When you're working with text or less well organized displays, you often note recurring patterns, themes, or "gestalts," which pull together many separate pieces of data. Something "jumps out" at you, suddenly makes sense. . . . The human mind finds patterns so

quickly and easily that it needs no how-to advice. Patterns just "happen," almost too quickly . . . [and so] need to be subjected to skepticism—your own or that of others—and to conceptual and empirical testing. (Miles & Huberman, 1994, p. 246)

Unless your colleagues or supervisors are unable to accept this, we urge that you write memos in the first person, firmly reporting "what I did" and "what I saw" in the data. This will help you later to avoid passively reporting that a theme "emerged" or a category "appeared." Whatever your chosen method, your study will benefit from your joining the increasing community of qualitative researchers who document, date, and account for their ideas so they can tell a category's story to doubting readers and track back how it has changed and been used. At the end of this chapter, we discuss ways to do this on the computer.

Managing Categories: Index Systems

Most qualitative research methods require some way of gathering and managing ideas and categories (and the definitions and memos that are attached to them). But few qualitative scholars offer advice on how to do it, and we have seen students tied up for months as they try to establish safe and flexible category management systems. Like any growing body of heterogeneous materials (correspondence, clothes, library books, recipe cards, addresses, tools, photographs, and more), research ideas will be hard to access, easily lost, and generally problematic if they are not stored logically. Our advice is that you manage the growing ideas from the start of the project rather than wait and then discover you are losing them. However you do it, you will be rewarded for developing a storage system early and reworking it as you go on. You can do this manually (with index cards or files) or on the computer. Most specialized qualitative research software now allows you to store categories and subcategories in "trees," which you can manage and view much as you manage and view the structure of your folders and files. A helpful storage system will incorporate what is known prior to the study; will be logical, so you can find ideas again; and will easily accommodate the emerging ideas you are about to create. (For help with manual index systems, see Lofland & Lofland, 1995; Miles & Huberman, 1994; for managing categories on the computer, see Richards, 2005; Richards & Richards, 1995.)

Models and Diagrams

Some methods, and within those some researchers, use modeling and displays to assist in the abstraction process. These techniques were given prominence with the publication in 1984 of the first edition of Miles and Huberman's "sourcebook," and the development since of computer imaging techniques that can relate to the items in a qualitative project. Note that *diagram, model, network,* and related terms have many meanings in this context. Qualitative researchers have always diagrammed, but usually in very impressionistic ways. (For ways of "seeing a whole," see Richards, 2005, chap. 9; for an extended example of the use of diagrams in grounded theory, see Strauss, 1987, chap. 8; for a very extended discussion of modes of display, conceptual frameworks, and models, see Miles & Huberman, 1994.)

⧗ UJING YOUR JOFTWARE FOR MANAGING IDEAJ

The primary difference between commercial database software and specialized qualitative software is that the latter is designed to help with the processes of analysis and abstraction. Software packages designed for qualitative research will store not only materials but also ideas, concepts, issues, questions, and theories.

The primary difference between using a computer system in qualitative research and using a manual system is that the computer gives a different sort of access, allowing for flexible growth in the webs of ideas the data are producing and allowing the researcher to manage those burgeoning ideas by storing them, defining them, accessing them, and writing about them.

The contribution software can make at the early stages of abstracting is considerable. All researchers are diffident about first ideas. If scribbled on stick-on labels, the ideas may be lost. The computer makes it easy for the researcher to store them, define them, write about and revisit them, and revise and review them as ideas build up.

Approaches

Your computer software will offer ways to make, manage, and develop categories and work with them from the early stages of your project. And all software offers some ways of asking questions about the relations between the categories you are reflecting on.

With your qualitative software, or other software for modeling, explore computer-based diagramming. This allows you to "play" with ideas in ways that are impossible with manual methods, including layering, labeled links, and live access to data.

Software will support the following processes, which you need to learn:

᠁ How to manage and move the categories for abstracting from your data

᠁ How to store definitions and how to describe and write memos about them and log changes

᠁ How to use search and query tools to explore the relations of categories

᠁ How to model your first ideas about the topic and the hunches growing from your exploration of the data

If you wish to try doing it in software, go to Appendix 1, where there is a guide to tutorials available online.

Advances

These are the areas where software has opened entirely new ways of working with qualitative data. As your project progresses, you will learn the uses of these tools to assist your exploration of the data, your checking of your hunches, and your developing of theories and reporting patterns and themes.

Category management is much more possible because software allows:

᠁ Flexibly copying, moving, and combining coding categories without losing coding

᠁ Linked logging of project process

Searching and querying of the data are now done in ways that were simply not considered previously. For example, software supports:

᠁ Creating (and optionally saving) questions about patterns of coding

᠁ Searching the text of documents

ℕ Expanding the results of searches to appropriate context

ℕ Automatically coding the results so they can become the basis for another question

ℕ Making matrices that demonstrate patterns and allow us to go to each cell to see what *those* people said about *this* issue

Models and diagrams were always part of qualitative research, but computer-based modeling offers many advances: for example:

ℕ Showing connections you made by coding or linking

ℕ Allowing the researcher to open data items from within the model to explore them further

Alerts

Just because these new tools are so exciting, be careful to use them thoughtfully and flexibly.

1. Cataloguing and ordering categories can become a passion! Stop when the catalog is good enough for your purposes. Tools for managing, reviewing, and moving categories and for developing ideas about their relationships should be used as indicated by your research goals and design. If unplanned, those processes dominate, since categories can be changed at any time, reordered at any time, and combined or deleted as the data direct your understanding of them. The challenge is to use this flexibility to assist abstraction while avoiding the trap of constantly reworking a workable index system (Richards, 2005, chap. 6).

2. Learn to conduct and interpret thoughtfully the powerful searches your software supports. When you come to use these tools in your own project, be careful always to prepare the question you are asking in plain language, and then interpret the result accurately. (For advice and cautions on using computer-based search tools, see Richards, 2005, chap. 8.)

3. Modeling with software offers attractive ways of "seeing" and showing your project. The risk is that you can "see" in a model a theme or pattern that is not reflected in your data. Always make the model true to the data.

SUMMARY

By now you should understand the processes of interpretation that accompany the clerical task of coding and why we code data. Coding is never an end in itself; it is a way of achieving categorization and interpretation, and these of course are sought and achieved in ways other than coding. In this chapter, we have explained how categories are discovered and used in analysis as well as the goal of concept development. We have looked at some of the ways of doing abstraction and at the sources and signs of abstract concepts. We have emphasized the need for management of ideas and development of categories and how these can be assisted by software. Categorization and conceptualization are processes that will enable you to identify patterns, explore hunches, draw in what you have learned from the literature to inform interpretation and explanation, and derive and justify your understanding. Doing these tasks rigorously is essential for high-quality qualitative research, and very exciting for the researcher.

RESOURCES

Carney, T. F. (1990). *Collaborative inquiry methodology.* Windsor, ON: University of Windsor, Division of Instructional Development.

Finch, J. (1986). *Research and policy: The uses of qualitative research in social and educational research.* London: Falmer.

Glaser, B. G., & Strauss, A. L. (1971). *Status passage.* Chicago: Aldine.

Lofland, J., & Lofland, L. H. (1995). *Analyzing social settings: A guide to qualitative observation and analysis* (3rd ed.). Belmont, CA: Wadsworth.

Miles, M. B., & Huberman, A. M. (1994). *Qualitative data analysis: An expanded sourcebook* (2nd ed.). Thousand Oaks, CA: Sage.

Mithaug, D. E. (2000). *Learning to theorize.* Thousand Oaks, CA: Sage.

Morse, J. M. (1994b). "Emerging from the data": The cognitive processes of analysis in qualitative inquiry. In J. M. Morse (Ed.), *Critical issues in qualitative research methods* (pp. 23–42). Thousand Oaks, CA: Sage.

Morse, J. M., Mitcham, C., & van der Steen, V. (1998). Compathy or physical empathy: Implications for the caregiver relationship. *Journal of Medical Humanities, 19*(1), 51–65.

Richards, L. (1990). *Nobody's home: Dreams and realities in a new suburb.* Melbourne: Oxford University Press.

Richards, L. (2005). *Handling qualitative data: A practical guide.* London: Sage.

Richards, T. J., & Richards, L. (1995). Using hierarchical categories in qualitative data analysis. In U. Kelle (Ed.), *Computer-aided qualitative data analysis: Theory, methods, and practice* (pp. 62–68). London: Sage.

Strauss, A. L. (1995). Notes on the nature and development of general theories. *Qualitative Inquiry, 1,* 7–18.

Turner, B. A. (1981). Some practical aspects of qualitative data analysis. *Quality and Quantity, 15,* 225–247.

Van den Hoonard, W. C. (1999). *Working with sensitizing concepts.* Thousand Oaks, CA: Sage.

8

Revisiting Methodological Congruence

n Chapter 3, we introduced the major types of qualitative methods, explaining that each method facilitates the achievement of particular analytic goals. Although all of the methods discussed use coding strategies and procedures for categorizing and/or theme-ing data (presented in Chapters 6 and 7), *how* the researcher approaches coding and categorizing differs from method to method. Each method demands that the researcher think about data in a particular way. Within any method, researchers may also incorporate different research strategies and ways of approaching and working with data. These strategies enrich the analysis and facilitate understanding. The analytic processes fit together in a coherent way to form the various qualitative methods, making them very different from each other. The modes of coding, categorizing, and theorizing are sensitive to the method used and appropriate additional analytic strategies.

In Chapter 3, we also sketched how coherence works in three methods: phenomenology, ethnography, and grounded theory. The differences in methods start from different assumptions, from the researcher's asking and answering different sorts of questions and using different types of data or data in different forms. From method to method, the researcher thinks differently, managing and exploring data by employing different analytic strategies. In this chapter, we revisit the same three methods, focusing mainly on data-making and analytic strategies. In the sections that follow, we discuss the strategies most commonly used in each method. Although it is not possible to describe the techniques in detail in this short chapter, we hope at least to provide adequate information for you to plan your proposal and to be aware of options and alternatives. If you need to use one of these strategies, the literature cited in the text as well as the resources listed at the end of the chapter will guide you to more information and examples.

〲 PHENOMENOLOGY

Phenomenology is an approach that enables the researcher to understand the meaning of phenomena. Phenomenological researchers often do not talk about a method per se (Wilson & Hutchinson, 1991), but rather discuss the tradition and the reading, reflection, and writing process that enables the researcher to transform the lived experience into a textual expression of its essence. Van Manen (1990) notes that "phenomenology differs from almost every other science in that it attempts to gain insightful descriptions of the way we experience the world pre-reflexively, without taxonomizing, classifying, or abstracting it" (p. 9). As discussed in Chapter 3, phenomenological reflection takes place within the four existentials: *temporality* (lived time), *spatiality* (lived space), *corporeality* (lived body), and *relationality* or *communality* (lived human relation). The end result is not effective theory that enables us to explain reality, but rather "plausible insights that bring us in more direct contact with the world" (van Manen, 1990, p. 9).

Data-Making Strategies

Introducing phenomenology in Chapter 3, we emphasized the idea of *bracketing* previous knowledge—that is, placing it aside. Bracketing is relevant to the way the researcher tackles each of the tasks discussed in the preceding chapters: research design, making data, coding, and abstracting. Bracketing—of one's theories, prior knowledge, and experience with the phenomenon—is intended to allow the investigator to encounter the phenomenon "freshly and describe it precisely as it is perceived" (Giorgi, 1997, p. 237). The researcher achieves bracketing by making these notions explicit, writing them down in a diary or memos. Ray (1994) notes that in phenomenological research, the research questions are not predetermined; rather, the researcher follows the cues of the participant and the conversation proceeds thoughtfully.

The purpose of this reflection is to find out what is essential in order for the phenomenon *to be*. To grasp an essence, the researcher reflects on concrete experience, trying to imagine it from all aspects. Other data sources enhance this period of reflection. Phenomenologists obtain descriptions from literature and poetry, explore instances of the phenomena of interest in movies and art, and reflect on the phenomena using the phenomenological literature as a lens.

Analytic Strategies

Phenomenological analysis is a process of reading, reflection, and writing and rewriting that enables the researcher to transform the lived experience into a textual expression of its essence (van Manen, 1990, p. 10). Researchers gain insights into the phenomena they study by using a number of strategies: tracing etymological sources, searching idiomatic phrases, obtaining experiential descriptions from participants, observing and reflecting further on the phenomenological literature, and writing and rewriting (Ray, 1994; van Manen, 1990).

Researchers select words or phrases that describe particular aspects of the lived experience they are studying and reflect on these. They may group and label similar expressions and eliminate expressions they believe are irrelevant. They then cluster and label groups of expressions that bear close relationships to one another and check this identified core of common elements against a selection of original descriptors obtained in conversations with participants. For the phenomenological researcher, the value of the process of writing and rewriting cannot be overestimated. Insight is developed through reflection, and researchers have found that discussing their texts with others is helpful.

Several authors have described the actual process of doing phenomenology as a series of steps. Although seeing the process laid out in steps allows the student to form an idea of how to proceed, the student should be aware that the process is not stepwise, not linear, but iterative. Labeling the different elements of the process as steps therefore tends to trivialize the cognitive work involved and actually moves the student away from thinking phenomenologically. With this warning, we describe several methods below.

Van Manen (1990) suggests that researchers use their own personal experience as a starting point. From there, they may *trace etymological sources* of the phenomena of interest and search for idiomatic phrases. They may obtain *experiential descriptions* from others, interview for personal life stories, observe to share experiential anecdotes, obtain experiential descriptions from the literature, and consult the phenomenological literature. They then conduct *thematic analysis,* write, and rewrite as a "measure of thoughtfulness" (van Manen, 1997).

Giorgi (1997) presents "five basic steps: (1) collection of verbal data; (2) reading of these data; (3) breaking data into some kind of parts; (4) organization and expression of data from a disciplinary perspective; and (5) [making a] synthesis and summary of the data for purposes of communication to the scholarly community" (p. 237).

Spiegelberg (1975; cited in Boyd, 1993) describes seven steps in arriving at essence. The first is *intuition,* which involves developing one's consciousness through looking and listening. This is followed by *analyzing,* which involves identifying the structure of the phenomenon under study and which occurs through *dialectic* (i.e., the conversation between participant and researcher). This knowledge is created through a joint project in which respondent and researcher are both committed to describing the phenomenon under study. The third phase is *describing the phenomenon;* however, premature description is one of the potential dangers in phenomenology. Description directs the listener to explore his or her own experience of the phenomenon. Insight becomes communicated through description. The next two steps involve *watching modes of appearing* and *exploring the phenomenon in consciousness*. At this stage, the researcher reflects on the relationships, or *structural affinities,* of the phenomenon. The final two stages are *suspending belief* (phenomenological reduction) and *interpreting concealed meanings.*

ETHNOGRAPHY

As discussed in Chapter 3, ethnography is based on the theory of culture, with the assumptions that cultural beliefs, values, and behaviors are learned, patterned, and may change. They may be overt (as ethnic identity is formed) or subconscious, below the level of awareness. The ethnographer, then, needs research designs and strategies for eliciting cultural beliefs and values that are implicit within the culture and strategies that enable the identification, comparison, and contrasting of those characteristics.

The first analytic goals of researchers in ethnography are to learn "what is going on" and to be comfortable in the setting. Researchers achieve these goals by completing first-level description, and then, once they have some understanding of the setting, working to obtain thick description. The strategies described below in each category are not fixed in stone, and each may be used for other analytic goals.

First-Level Description

First-level description is the description of the scene, the parameters/boundaries of the research group, and the overt characteristics of group members; it may include demographic characteristics, the history of the populations, maps of relevant areas, and so forth. First-level description places the reader in the context of the study and sets the stage by providing background information for the subsequent analytic reporting. These data are usually a synthesis of public documents and other publicly available information along with basic description of the setting through the use of mapping, photography, or other means to illustrate the setting and describe the people. First-level description is usually presented in narrative form, supplemented by tables, maps, or photographs as necessary.

Strategies to Enhance First-Level Description

Many strategies for orienting to the setting are described in the literature. "Doing something" can ease the researcher's awkwardness upon entering a setting and help the researcher to get to know the participants. Initial strategies include the following:

⚝ Maps/floor plans of the setting or simple line drawings can serve as "memory joggers." Photographs are a quick and effective means for recording the scene.

⚝ Organizational charts can help the researcher to learn the formal and informal relationships and chain of command.

⚝ Documents frequently give important clues to the history of the setting; initially, the researcher may simply want to make a list of the kinds of documents available.

⚝ Documenting routines helps to give the researcher a feel for the rhythm of the setting. The holistic approach of ethnography is aided by the researcher's inclusion of all routines. For instance, if the setting is a hospital, the researcher should include the routines observed on weekends and nights; if the setting is a community, the researcher should include routines observed in all seasons as well as on special days and holidays; if the setting is an institution, observations should include rush hours as well as slack times.

Other strategies help the researcher to assess the status quo:

⧵ A rapid community survey allows teams of researchers to grasp at least the more urgent problems.

⧵ Interviews with key community members, or focus groups, may help to highlight issues, allowing the researcher to get a feel for the research domain, obtain relevant items for a questionnaire, or provide a rapid form of evaluation.

⧵ The researcher may study pertinent records and newspaper coverage to supplement and inform participant observation.

The literature also suggests strategies for documenting interpersonal relationships. Researchers interested in informal relations may employ sociograms, network diagrams, and drawings. Many techniques used in social network analysis can help researchers to see relations as they are seen by participants. More formal relations may be portrayed in kinship charts and organizational charts.

Thick Description

Researchers develop thick description (see Geertz, 1973) by processing interviews and field notes in which informal conversations and observations are reported and through theoretical insights developed from these materials. Researchers condense these materials by summarizing, synthesizing, and extracting the essential features or characteristics of the situation. To be able to synthesize without losing any significant detail or variation requires skill. There must be consistency between one's observations, reports from the participants, and information in the literature if the study is to achieve generalization, validity, and abstraction. For example, in a study of young mothers in neighborhoods, a researcher might describe how they watch out for each other during the day. Such a description of one case in one neighborhood is context bound. But if the researcher's observations and participants' accounts are placed within the theoretical body of social support literature, the study can be tried against other situations, and it moves from description to analysis.

Strategies to Enhance Thick Description

All ethnography seeks thick description, and in all ethnography thick description is built using three strategies for making data: observations, interviews, and diaries. Revisit Chapter 5 to consider why these are central to ethnography.

Ethnographers also often use quantitative surveys early in their studies to establish the base data, or later in the ethnographic fieldwork, when they have identified and delineated particular phenomena and need to know how those phenomena are distributed among the groups they are studying. In her ethnographic study on the cultural responsiveness to pain, Morse (1989a) used survey methods and Thurstone's paired comparison technique (Nunnally, 1978) and found that how painful the participants perceived certain conditions to be was culturally taught and varied among cultures. Richards (1990) used statistical analysis of survey results to indicate clusters of neighboring behavior that assisted the development of typologies from the qualitative data. Quantitative measures may supplement qualitative: Margaret Mead, for instance, often used psychometric tests as a part of her traditional ethnography.

Comparison

Cultural values are not easily studied or directly reported by participants, so researchers must obtain indicators of such values indirectly.

Strategies to Enhance Comparison

Identification of cultural values and perspectives may be enhanced through the direct comparison of two cultures or through "shadow comparison"—that is, comparison of the results from a given cultural group with those described in the literature or with knowledge of the researcher's own culture.

⦉ Comparison of two cultural groups that are very different can provide information on as wide a range of dimensions as possible.

⦉ Before-and-after research design is often less than satisfactory in qualitative work if it requires "blinding" the researcher, collecting baseline data, implementing interventions, and then collecting data to demonstrate the changes that have occurred.

⦉ Asking comparative questions can be the key to ethnographic analysis. It involves a stance that constantly asks, "What are the characteristics of . . . ?" "What are the types of . . . ?" and "How does . . . relate to . . . ?"

⦉ Researchers may investigate the implicit structural features of concepts and phenomena by using techniques of ethnoscience (see Spradley, 1979; Werner & Schoepfle, 1987a, 1987b).

Researchers may establish commonalities among different types of components through the use of dyadic and triadic card sorts. In this technique, the researcher writes all of the attributes of each component on cards, one attribute per card; the interviewer presents the full set of cards to the participant and asks him or her to sort the cards into piles (two piles in a dyadic sort, three in a triadic sort) and then name each pile and describe how the piles are different. In a Q-sort, participants are given the cards to sort into piles, but they are free to decide how many piles to make. The participants' dominant piles determine the segregates (major categories) and subsegregates. In this way, the researcher is able to sort and display the categories as a taxonomy. The researcher then needs to search further to determine whether the characteristics are unique to each category and whether further characteristics related to each classification can be identified. Figure 8.1 shows results displayed in several formats, using part of a taxonomy from a study conducted by Morse (1989b). The box displays various levels as segregates and subsegregates.

Results

During analysis, the segments of data may be considered as pieces of a puzzle that fit together to give a complete, holistic, thick, and rich description of the cultural perspective on the research problem. Because the researcher is trying to gain understanding, ethnographic research is inherently flexible: The researcher encounters something that does not fit, or does not immediately make sense, and then identifies a way to

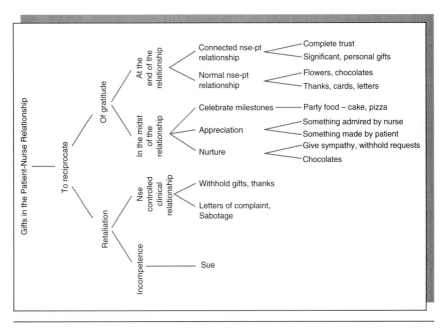

Figure 8.1   Taxonomic Structure of a Box Diagram

analyze that particular situation. Thus ethnographic research design is malleable but, like any method, only within limits. (Don't forget that if you add new strategies, you will need to notify and obtain permission from your ethics review board.)

Ethnographies are usually presented as giving comprehensive, consistent, and logical portrayals of the groups or phenomena studied (from the emic perspective). Because of the depth of description required and the scope of many ethnographies, they are often presented as monographs rather than articles. They may present particular challenges in reporting, because the richer the data and the more varied the portrayals, the more identifiable participants may be, however careful the researchers have been in changing names and details. You should consider these challenges in designing your research.

⟩⟩ GROUNDED THEORY

Recall that grounded theory is a method based on symbolic interactionism, used to explore process. The end result is usually midrange theory specific to a particular process and a particular situation.

Data-Making Strategies

Because grounded theory is about understanding process and theoretically constructing models based on the stages and phases of phenomena over time, the researcher's stance and approaches to data must maximize the recording of the process. Frequently these data are interviews or observations collected prospectively or narratives collected retrospectively— stories reporting retrospectively about what happened and about the changes that evolved. Data documents may also include field notes from observations.

But whatever the ways of making data, of particular importance in grounded theory are the researcher's memos and other written documentation of insights gleaned during the process of analysis. As theory is constructed, "one can frequently sense the hovering presence of memos which arise out of codes and ideas generated" (Strauss, 1987, p. 109). There is a strong emphasis on dialogue and the challenging of ideas and their development as data accrue. The lone researcher or the team encapsulates this in memos, "a running record of insights, hunches, hypotheses, discussions about the implications of codes, additional thoughts" (Strauss, 1987, p. 110).

Data Preparation

Analysis begins with the first exploration of the topic and literature and is ongoing throughout the study. Hence it is crucial that the researcher not process data records in bulk and only then consider them. Often a grounded theory study will begin with memos, and then ideas grow from the first field notes of interviews. As researchers explore similar incidents, events, and experiences that recur and begin to develop concepts, they revisit memos written earlier.

Records of interviews or field notes are usually transcribed, and these transcripts are checked for accuracy. At this time, descriptive codes (such as codes for demographic characteristics of the participant) are entered and linked to each transcription. But full transcription is a matter of contention in the literature. Glaser (1998) argues against taping and transcribing on the grounds that this prevents theoretical sensitivity. Conversely, we share a concern that the researcher who does not transcribe relies too much on memory and loses contact with the richly detailed text. Participants' words may be a crucial source of "in vivo" codes and provide a vivid illustration from data that are rich, accurate, and verifiable. Field notes and the researcher's diary should normally also be entered into the computer and treated as data for analysis.

Analytic Strategies

All the main analytic strategies used in grounded theory are comparative: open coding, memoing, categorizing, and the integration of data through diagramming.

Strategies That Facilitate the Identification of Process

Two strategies are particularly useful for facilitating the identification of process:

⚜ The researcher constructs a timeline for each participant, sequencing the major events, emotional responses, or strategies described in the data, and then places the timelines one under the other so that certain events are aligned rather than in months or other calendar designations. This allows the researcher to use the timelines to compare individual cases.

⚜ The researcher identifies the course or *trajectory* of each participant's story and compares and contrasts major events. Although the stories may not be identical, and the participants may not use the same labels for important events, this comparison should make it possible for the researcher to identify common patterns. The researcher then examines data for events in common among participants and similar emotional responses or behavioral strategies of responding to those events. From this comparison, the researcher can develop a rich description of these events and look for common antecedents and consequences.

Strategies for Coding

The transcription and analysis of field notes or interviews begins immediately following the events, and there is a continuous and responsive interaction between the collection of data and analysis, with the data directing the coding process and vice versa. In grounded theory, in vivo codes are used alongside, and often in preference to, what Strauss (1987) calls "sociological codes"—that is, such known concepts as social support or coping. The labels for codes are taken directly from the language that participants themselves used. When new categories are created, they are often given new names.

Glaser (1978) recommends that to increase their repertoires of codes (and to prevent them from becoming stuck and overusing "pet" codes), researchers should use families of theoretical codes. For instance, Glaser writes of the "six Cs": causes, contexts, contingencies, consequences, covariances, and conditions. Attention to such families of codes will facilitate coding and the researcher's understanding of the richness of theory.

Both Strauss (1987) and Glaser (1978) suggest some rules to guide coding. In Chapter 3, we recommended thoughtful review of the differences in coding style developed between these two founders of grounded theory. We suggest that you read critically the accounts from both authors to help you understand the processes of coding. Both, in different ways, recommend that the researcher approach coding not as tagging of text by categories but as a way of "opening up" the text to explore its meanings.

⟡ The researcher asks a set of questions that will keep the focus on the purpose of the study, such as the following: What are these data a study of? What category is this instance an example of? What is actually happening in these data? What are the *dimensions* of this category? Under what *conditions*, and with what *consequences*, will this occur?

⟡ The researcher examines and codes transcripts line by line, highlighting important passages and creating theoretical memos (noting insights, comparisons, summaries, and questions).

⟡ The researcher gathers passages or significant portions of the text as categories, bringing them into comparative view with similar text.

⟡ The researcher uses *constant comparison* in comparing indicator with indicator, concept with concept. This enables the researcher to identify patterns and thus to label similar incidents as a category and to identify the properties of the category.

⟡ The researcher conducts *theoretical coding* to follow the "lead" of the data, seeking other instances and related instances to increase the degree of abstraction of the analysis.

〰 Once the researcher has verified the initial data in other instances and has identified interchangeable examples, *saturation* is reached.

〰 In the final step of the coding process, the researcher identifies the linkages connecting the various categories by comparing and contrasting the conditions and consequences of the relationships among categories.

Strategies With Memos

The key component of theory development is the writing of theoretical memos—which are always revisited, constantly growing, and always treated as data. While coding, the researcher records in memos information, ideas, insights, thoughts, and feelings about the relationships in the emerging theory. It should be noted that researchers may write memos about other memos. The researcher sorts and compares memos as the theory becomes more streamlined. If, at any stage, there is a question as to whether personal biases are affecting perceptions of the phenomenon, the researcher should augment observations with informal interviews to clarify perceptions and "ground" the data.

Theory-Building Strategies

As coding and memoing continue, the researcher codes the text in categories and is able to label these categories, alter the memos about them, identify core categories, and note similarities and differences between categories. Transitions or turning points in the data mark the margins between stages. Identifying characteristics, and noting the presence or absence of these primary characteristics, helps the researcher to recognize the underlying rationales for participants' chosen decisions and substantially aids the theoretical development of the study.

There is constant interaction among sampling, data collection, the emerging analysis, and theory construction. This process of analysis is highly interactive, with the researcher going back and forth between the emerging conceptual scheme or theory and data sorts. As new ideas are identified, continual data making is redirected to elicit further information regarding these ideas. As explained in Chapter 5, during these next stages the researcher probes for more information, or missing data, to develop the theory or repair gaps and address unanswered questions.

The researcher may seek additional negative (i.e., variant) cases or unique cases until these data are saturated and built into the emerging theory. At this point, the researcher seeks confirmation of a growing theory or a specific hypothesis. If the information gleaned from the data is unexpected or unanticipated, it may be necessary to change direction and follow the new leads by interviewing a different group of participants.

As the researcher continues to code, there is active seeking of "integration," the stage of comprehension, an in-depth knowledge of the data, and synthesis—the ability to report, or to be able to tell a "generalized story": "These people do this and that." Once the stage of synthesis has been reached, theory development begins. At this point, *critical junctures* may be identified. The researcher now becomes more focused, filling in gaps and areas that are thin and coding selectively rather than coding all that once appeared relevant.

In both Glaser's and Strauss's accounts of grounded theory method, there is emphasis on seeking a core category around which local theory will be built. The two describe this differently—but don't allow the dispute to distract you from what is central in the method. In both accounts, the researcher seeks a core category. In gaining a greater understanding of the research topic, the researcher may be able to identify the primary theme— the core category or the basic social process (BSP). As Glaser (1978) describes it, the researcher finds BSPs by "looking" (p. 94) for a main theme while coding (by discovery) or for an emergent fit (p. 107). BSPs may be derived through a theoretical code, a process, a condition, two or more dimensions, a consequence, and so forth. Central to the analysis, they recur frequently in the data and vary with conditions and consequences. Because they are "scattered through many categories," BSPs will be slower to become saturated than other categories, and will have theoretical "grab" (see Glaser, 1978, pp. 94–107).

Strauss (1987) writes of seeking a core category that brings together understanding of the data. He quotes Glaser's description of the core category:

> It is relevant and works. Most other categories and their properties are related to it, which makes it subject to much qualification and modification. In addition, through these relations among categories and their properties, it has the prime function of *integrating* the theory and rendering it *dense* and *saturated* as the relationships are discovered. These functions then lead to theoretical *completeness*— accounting for as much variation in a pattern of behavior with as few concepts as possible, thereby maximizing parsimony and scope. (Glaser, 1978, p. 35)

Once this core is identified, sampling and coding become more targeted and focused—a process known as *selective coding*. The researcher uses diagramming and mapping extensively to facilitate the analytic process of delineating stages and the characteristics of each stage. These processes enable the researcher to attain an increased level of abstraction and clarify the development of theory.

One strategy of diagramming is the construction of typologies (Glaser, 1978). At its simplest, the researcher identifies two variables or emerging concepts that appear to contribute to the variants of the phenomenon of interest and, using a two-by-two matrix, explores the effects of the presence or absence of each variable in combinations.

☒ SUMMARY

By this stage of the project, you should be developing a solid understanding of doing qualitative research. You are aware of how important it is to *think qualitatively,* and that each qualitative method requires its own way of thinking and uses a particular perspective and specific strategies to reach its analytic goals. The product of each method, in turn, provides a perspective on reality that is specific to that method.

You understand by this stage that the congruence of the method you have selected provides distinct perspectives and produces distinct products. The major methods described in detail in this chapter—phenomenology, ethnography, and grounded theory—provide methodological toolboxes for eliciting meaning, thick description, and process.

☒ RESOURCES

Phenomenology

Giorgi, A. (Ed.). (1985). *Phenomenology and psychological research*. Pittsburgh, PA: Duquesne University Press.

Moustakas, C. (1994). *Phenomenological research methods*. Thousand Oaks, CA: Sage.

Munhall, P. L. (1994). *Revisioning phenomenology* (Publication No. 41–2545). New York: National League for Nursing Press.

Spiegelberg, H. (1975). *Doing phenomenology: Essays on and in phenomenology*. The Hague: Martinus Nijhoff.

van Manen, M. (1990). *Researching lived experience: Human science for an action sensitive pedagogy.* London, ON: Althouse.

van Manen, M. (1997). From meaning to method. *Qualitative Health Research, 7,* 345–369.

Ethnography

Agar, M. H. (1996). *The professional stranger: An informal introduction to ethnography* (2nd ed.). San Diego, CA: Academic Press.

Alasuutari, P. (1995). *Researching culture: Qualitqtive method and cultural studies.* London: Sage.

Denzin, N. K. (1997). *Interpretive ethnography: Ethnographic practices for the 21st century.* Thousand Oaks, CA: Sage.

Fetterman, D. M. (1998). *Ethnography* (2nd ed.). Thousand Oaks, CA: Sage.

Geertz, C. (1973). *The interpretation of cultures: Selected essays.* New York: Basic Books.

Muecke, M. A. (1994). On the evaluation of ethnographies. In J. M. Morse (Ed.), *Critical issues in qualitative research methods* (pp. 187–209). Thousand Oaks, CA: Sage.

Schensul, J. J., & LeCompte, M. D. (Series Eds.). (1999). *Ethnographer's toolkit* (7 vols.). Walnut Creek, CA: AltaMira.

    The seven volumes in this set of books are as follows: Volume 1, Schensul, J. J., & LeCompte, M. D., *Designing and conducting ethnographic research;* Volume 2, Schensul, S. L., Schensul, J. J., & LeCompte, M. D., *Essential ethnographic methods;* Volume 3, Schensul, J. J., LeCompte, M. D., Nastasi, B. K., & Schensul, S. L., *Enhanced ethnographic methods;* Volume 4, Schensul, J. J., LeCompte, M. D., Trotter, R. T., II, Cromley, E. K., & Singer, M., *Mapping social networks, spatial data, and hidden populations;* Volume 5, LeCompte, M. D., & Schensul, J. J., *Analyzing and interpreting ethnographic data;* Volume 6, LeCompte, M. D., Schensul, J. J., Weeks, M. R., & Singer, M., *Researcher roles and research partnerships;* Volume 7, Schensul, J. J., LeCompte, M. D., Hess, G. A., Nastasi, B. K., Berg, M. J., Williamson, L., Brecher, J., & Glasser, R., *Using ethnographic data.*

Van Maanen, J. (1988). *Tales of the field: On writing ethnography.* Chicago: University of Chicago Press.

Werner, O., & Schoepfle, G. M. (1987a). *Systematic fieldwork: Vol. 1. Foundations of ethnography and interviewing.* Newbury Park, CA: Sage.

Werner, O., & Schoepfle, G. M. (1987b). *Systematic fieldwork: Vol. 2. Ethnographic analysis and data management.* Newbury Park, CA: Sage.

Whyte, W. F. (Ed.). (1991). *Participatory action research.* London: Sage.

Wolcott, H. F. (1995). *The art of fieldwork.* Thousand Oaks, CA: Sage.

Wolcott, H. F. (1999). *Ethnography: A way of seeing.* Walnut Creek, CA: AltaMira.

Grounded Theory

Chenitz, C., & Swanson, J. M. (1986). *From practice to grounded theory.* Reading, MA: Addison-Wesley.

Dey, I. (1999). *Grounding grounded theory: Guidelines for qualitative inquiry.* New York: Academic Press.

Glaser, B. G. (1978). *Theoretical sensitivity: Advances in the methodology of grounded theory.* Mill Valley, CA: Sociology Press.

Glaser, B. G. (Ed.). (1996). *Gerund grounded theory: The basic social process dissertation.* Mill Valley, CA: Sociology Press.

Glaser, B. G. (1998). *Doing grounded theory: Issues and discussions.* Mill Valley, CA: Sociology Press.

Glaser, B. G., & Strauss, A. L. (1967). *The discovery of grounded theory: Strategies for qualitative research.* New York: Aldine.

Strauss, A. L. (1987). *Qualitative analysis for social scientists.* New York: Cambridge University Press.

Strauss, A. L., & Corbin, J. (Eds.). (1997). *Grounded theory in practice.* Thousand Oaks, CA: Sage.

Strauss, A. L., & Corbin, J. (1998). *Basics of qualitative research: Techniques and procedures for developing grounded theory* (2nd ed.). Thousand Oaks, CA: Sage.

Wilson, H. S., & Hutchinson, S. A. (1996). Methodologic mistakes in grounded theory. *Nursing Research, 45*(2), 122–124.

Part III

GETTING IT RIGHT

9

On Getting It Right and Knowing If It's Wrong

What makes a study solid? How can a researcher convince the readers of the research report of its rigor? How can a researcher demonstrate and communicate its credibility? In this chapter, we discuss criteria for ensuring and establishing reliability and validity in qualitative research, as well as criteria for evaluating a study. We cover strategies the researcher should use when planning the project, while conducting the study, and while bringing the study to completion. We also look at processes and procedures that continue to establish the legitimacy of the study after the first study has been completed.

Consider the terms we have selected here: *reliability* and *validity*. As discussed in Chapter 5, some authors have argued that these terms have no place in qualitative inquiry (e.g., Lincoln & Guba, 1985). Rather, they insist that because qualitative inquiry is subjective, interpretive, and time and context bound, "truth" is relative and "facts" depend upon individual perceptions. Therefore, they argue, *reliability* and *validity* are terms that belong to the positivist paradigm, and qualitative researchers should use different terminology. Others have argued that in the qualitative context, criteria for reliability and validity must be different from those used in a quantitative context (e.g., Leininger, 1994). In 1985, Lincoln and Guba recommended that qualitative researchers substitute for the concepts of reliability and validity the following aspects of *trustworthiness: truth value*, which is the *credibility* of the inquiry; *applicability*, which is the *transferability* of the results; and *consistency*, which is the *dependability* of the results.

We have no quarrel with the suggestion that reliability and validity are determined differently in qualitative inquiry than in quantitative work, but we regard it as essential that determining reliability and validity remains the qualitative researcher's goal (see also Kvale, 1989, 1995; Maxwell, 1992; Sparkes, 2001). Briefly, reliability requires that the same results would be obtained if the study were replicated, and validity requires that the results accurately reflect the phenomenon studied.

As one reads the qualitative literature, it becomes evident why these terms have been seen as problematic. Replicating a qualitative study is sometimes impossible and always difficult, because the data are richly within the particular context (Sandelowski, 1993). Attempting to judge the representation of "reality" is highly problematic, because qualitative researchers always view social reality as a social construction (Altheide & Johnson, 1994). However, to claim that reliability and validity have no place in qualitative inquiry is to place the entire paradigm under suspicion; such a claim has ramifications that qualitative inquiry cannot afford, and it diverts attention from the task of establishing useful and usable measures in the qualitative context. Qualitative researchers can and do defend their own work as solid, stable, and correct. It is these claims that give qualitative research legitimacy and thus the right to be funded, to contribute to knowledge, to be included in curricula, and, most important, to inform policy and practice. In this chapter, we offer some ways in which qualitative researchers can ensure, attain, and communicate reliability and validity.

※ ENJURING RIGOR IN THE DEJIGN PHAJE

Appropriate Preparation (Skill/Knowledge Level) of the Researcher

Any study (qualitative or quantitative) is only as good as the researcher. In qualitative research, this is particularly so because *the researcher is the instrument*. The researcher's skills ensure the quality and scope of data, the interpretation of the results, and the creation of the theory. Therefore, the onus is on the researcher to be maximally prepared in qualitative methods *before* beginning a project. That is why we wrote this book. Research methods are the tools of the researcher, and there is a best (and easiest) way to attack any problem. Methods are continuously being refined and improved, and learning methods is an art that does not stop with the completion of the first project. Learning new methods, developing strategies, improving techniques, and developing skills as an observer, an interviewer, and an analyst are lifelong commitments.

Appropriate Review of the Literature

Surveying the literature allows you to get a grip on what is known and to learn where the holes or weak areas are in the current body of knowledge. New theory never simply renames what has been described and reported previously. You must know what is already known, and when you see it, you must recognize, acknowledge, and give appropriate credit to it. In short, when you are in the field, it is essential that you be able to recognize a previously reported concept or pattern, refer to established explanations and theories, and recognize any variations between what was previously discovered and what you are now seeing.

In earlier chapters, we stressed the different ways in which different methods combine prior knowledge, knowledge gained from the data, and the testing of both against discovery. How you do this in your project will depend on the method you have chosen. All methods, however, require that the researcher balance the use of what is already known with the discovery from the data. The interplay of prior knowledge and discovery is critical in the process of making and developing concepts. "Starting with them (deductively) or getting gradually to them (inductively) are both possible. In the life of a conceptualization, we need both approaches" (Miles & Huberman, 1994, p. 17; for a discussion of deduction and induction in grounded theory, where prior knowledge is often assumed to be irrelevant, see Strauss, 1987, pp. 11–14).

Such knowledge and background do not, however, mean that you go looking for particular information to fit expectations. A common source of invalidity is the researcher's seeking out what the literature suggests he or she should find. Thus, whatever your method, you must bracket and put aside information you have learned about the phenomenon from the literature so that you can learn from the data. Once you have identified and developed what is being observed, you will be able to move the model reported in the literature over it, like a template, and compare the two.

In Chapter 3, we introduced the concept of bracketing as used in phenomenology. Whatever your method, it is a central process. Bracketing means putting your personal knowledge and the knowledge you have gained from the literature aside or making it overt by writing it down, so that you can see the research problem, the setting, and the data with fresh eyes and work inductively, creating understanding from data. Several strategies assist with bracketing. One is to write in your diary, before starting the project, your personal assumptions about what is going on, possible relationships, and what you think you will find. (Tutorial 1 suggests starting with a project document.) Making the implicit explicit brings your assumptions out into the open and helps you contain them. Another strategy is to summarize the information you have gathered in your reading to form a literature review, building an argument for you to conduct your study. Then set it aside—but do not put it away. Keep this literature in your mind. Pit it constantly against what you see. If you are confident it is right, use it as a springboard from which to start your inquiry. If you suspect it is incorrect, your constant reviewing and re-reviewing of the literature will allow you to compare and contrast your emerging data with the previous knowledge. You will need to update and discuss that review regularly and return to it at the end of your study, for all knowledge must be placed within the context of what is already known. Knowledge builds incrementally or by processes of replacement.

Thinking Qualitatively, Working Inductively

The key to rigorous qualitative inquiry is the researcher's ability to think qualitatively. This is an exhausting process of being constantly aware and constantly asking analytic questions of data, which, in turn, constantly address the questions asked. Qualitative inquiry constantly challenges assumptions; constantly challenges the obvious; reveals the hidden and the overt, the implicit, and the taken-for-granted; and shows these in a new light. Without such an active mode of inquiry, you risk a shallow, descriptive study, with few surprises, reporting the obvious. Is that a criterion of invalidity? Neglecting to pursue the real issues in a research topic will result in a weak study—probably one that will not be interesting enough to warrant publication. Yes, missing the essence of phenomena is a fatal validity issue.

Qualitative research does not, and should not, use a rigid prior conceptual framework that dictates the nature of the variables to be collected and the relationships among those variables. Such a framework would naturally guide the interviews, so that the researcher would deliberately seek data to fit the framework, showing that whatever the researcher planned to find was found, neat and tidy. Proceeding in such a manner invalidates the study. But there is a difference between using a rigid prior conceptual framework that dictates data collection and adhering to the philosophical basis of the study. Such philosophical paradigms as feminism, postmodernism, and critical theory do not create variable lists; rather, they provide ways of looking and focus the method used in a particular direction, for a particular purpose. Overlaying a study with such a theory, provided it is explicit, is not creating a source of invalidity but rather a lens to give your project a particular focus.

Let's add one other aspect to this difficult topic. Perhaps you are using nonparticipant observation because it is not possible to use interview data or other verbal data that would give you clues to what is actually going on—for instance, when observing neonates or elderly persons with Alzheimer's disease. In such a case, you need an interpretive framework to transform data into meaningful units. But this is a risk only if other data are not accessible to help you interpret your observations. To return to our supposed study of arrivals and departures (Chapter 2), we could conduct a microanalytic study of types of passengers arriving at the airport. Working inductively without a framework, we could describe the behaviors of the passengers and those waiting by enumerating precategorized movements of touch or nontouch, without attributing any meaning to them at all. The result would be a very insignificant study and a very boring read. But if we put the study in a framework of types of attachments according to roles and relationships, it becomes interesting and significant. The bottom line, therefore, is that you must be aware of when it is necessary to use an a priori framework, know the risk involved with using it, and be prepared to be wrong.

Using Appropriate Methods and Design

We have described, in earlier chapters, how to use an armchair walk-through to identify the most appropriate method to answer a particular research question. Is using a less appropriate method a source of invalidity? Perhaps.

You can drive in a nail with a hammer, a stone, or the heel of your shoe. The goal (of getting the nail in) is achieved no matter which tool you use, albeit clumsily if you use your shoe and awkwardly if you use a stone. But perhaps when you drive the nail, unexpected findings occur— the stone bends the nail or your shoe is damaged. Similarly, you may introduce sources of invalidity into your study by using a less-than-optimal research method. The golden rule of respecting methodological cohesiveness—of ensuring the best fit of the research question with the assumptions, strategies, types of data, and analysis techniques—ensures maximal validity. Remember that the different methods have different agendas; they are different ways of thinking about the data, and they produce different results. If you mix and match methods as the study proceeds, you may introduce ways of analysis inappropriate for your data, or data inappropriate to your question.

Validity requires accurate representation of what you are studying (Maxwell, 1992). When you are preparing to do the research, you should design the study to avoid redirecting it away from that goal. For instance, plan to use techniques of theoretical sampling, and plan ways to develop the sample to attain saturation. Plan how you may identify negative cases and seek additional cases. Although such plans are only plans (that is, they may not be followed to the letter), they at least alert the researcher, from the beginning, to the importance of covering the scope of the question, and they help to identify possible solutions that will enhance validity.

Proposals are a necessary and essential means of preparing the researcher for doing the project. The armchair walkthrough that is necessary for developing the proposal prepares the researcher for the project. The proposal may be used to obtain funding for the project, and it informs ethics review boards and the agency or institution in which data are being collected what the researcher is planning to do. We provide advice on this process in Chapter 11.

※ ENJURING RIGOR WHILE CONDUCTING A PROJECT

The most important strategies to enhance and maintain rigor take place during the actual conduct of the study itself (see Meadows & Morse, 2001). Whatever your qualitative method, the major thrust or direction of inquiry must be inductive, and this is crucial for the validity of the study. This is not to say that deduction is never used in qualitative inquiry. Deduction is used to develop conjectures and to verify data, but the dominant thrust should be inductive.

Asking *analytic questions* of the data is the key to solid and significant research. The active approach to inquiry is driven by the identification of questions that force the researcher to think, to confirm and to refute, to collect more data, and to pursue every avenue.

Using Appropriate Sampling Techniques

Sampling is the key to good qualitative inquiry and also to understanding the dilemmas of qualitative validity. Just as fishermen cast their lines into likely fishing holes, rather than randomly select places to fish, so qualitative researchers deliberately select participants for their studies. Qualitative researchers may seek bias, deliberately choosing the worst case or the best instance of an event rather than the average experience, for the characteristics of a phenomenon are more easily explored in the outstandingly good or bad examples. Average experiences are difficult to explore, as the characteristics of the phenomenon of interest are diluted and mixed in with other characteristics from other experiences.

Therefore, rather than employing random sampling, qualitative researchers seek valid representation with sampling techniques such as the following:

〰 *Purposeful sampling,* in which the investigator selects participants because of their characteristics (*Good informants/participants* are those who know the information required, are willing to reflect on the phenomena of interest, have the time, and are willing to participate; Spradley, 1979.)

〰 *Nominated or snowball sampling,* in which participants already in the study recommend other persons to be invited to participate

〰 *Convenience sampling,* in which those invited to participate in the study are simply those who are available to the researcher

〰 *Theoretical sampling,* in which the researcher deliberately seeks out persons to be invited to participate according to the emerging theoretical scheme (Morse & Field, 1995)

Occasionally, one does see a qualitative project reported in which a random sample has been used; for instance, a researcher might use semi-structured interviews if he or she has a large pool of participants and not enough information to use another type of interview. Consider the advantages and disadvantages of each way of sampling for your study and the threats of each to validity.

Responsiveness to Strategies That Are Not Working

Sometimes, when you are in the midst of data gathering, the data obtained do not appear as useful as they should be. Data analysis appears to be going nowhere—perhaps the material is not as detailed or as rich or as informative as it should be. When this happens, the response of many inexperienced researchers is to collect more data. But merely continuing to gather more poor data will not usually solve the problem. Rather, more of the same data will dishearten the researcher, putting him or her at risk of running out of funding or becoming more bogged down in analysis. The solution is to step back and consider why the data are not fruitful. It may be necessary to change data-making strategies, to select a different group of participants, or to observe or interview participants at different times. This problem particularly affects designs with homogeneous data. It may be necessary to change an interviewing project to observational strategies or to use a different approach to interviewing. The important thing is to stop using strategies that are not working. (Remember: If you do change strategies, you must notify those to whom you are responsible for the project design—partners in an action project, team members, supervisor, or thesis committee—and get any relevant ethics permission.)

Appropriate Pacing of the Project

By *pacing of the project*, we mean the synchrony of data making and analysis, the change of research techniques from obtaining information to verification as the project progresses and the change of the investigator's conceptualizing process from synthesizing to theorizing. Moving too quickly through a project, or moving to the next analytic task before the first analytic task is completed, will leave the research open to incomplete work, missed analytic opportunities, and possibly to having all parts of the project drawn together. Such a study loses validity, credibility, and rigor.

Related to this is the *assessment of saturation*. Data gathering must continue until each category is rich and thick, and until it replicates. It is saturation that provides the researcher with certainty and confidence that the analysis is strong and the conclusions will be right. If saturation is achieved, *negative cases*—instances that do not fit the emerging model— will be explored, and using techniques of theoretical sampling, the researcher will deliberately seek other similar negative cases until those categories are also saturated.

How will you know when a data set is saturated? Go back to the literature; read reports of studies that have used your chosen method and look for the ways a *satisfying* explanation is expressed. When data offer no new direction, no new questions, there is no need to sample further—the account satisfies. Often the first sign is that the investigator has a sense of having heard or seen it all. When data are saturated, events do not remain as a single instance; they have been replicated at least in several cases, and with that replication comes verification. When writing about the study, the researcher can fully describe not only the phenomenon but also the antecedents and the consequences and the various forms in which the phenomenon occurs.

Coding Reliably

In Chapter 6, we discussed the issue of reliability of coders across time and across teams, and the issues to be considered when reliability is sought for qualitative coding (Richards, 2005, pp. 98–99). It should by now be evident that the relevance of reliability of coding will depend on the methods used. In semistructured interviews, participants are all asked the same questions in the same order in a search for patterns of answers, so consistency in coding is important. The researcher must develop and record definitions used for particular codes, and if more than one person is coding, the researcher must attend to interrater reliability (i.e., consistency between coders). If using the computer, the researcher can do this rigorously, comparing the patterns of coding in each document in fine detail. By contrast, when a researcher uses unstructured, interactive interviews, imposing a consistent coding scheme can instead be a source of *invalidity*. If the researcher is learning about the phenomenon as the study progresses, interview content changes as the researcher becomes more informed about the phenomenon. As well, the domain narrows or broadens in scope as the researcher gains understanding and increasingly seeks more detail. Consistency works against the development of a rich interpretive study, the latter being the very purpose for which the researcher decided to use unstructured, interactive interviews. Rather than consistency, it is malleability that keeps the coding valid. The key is to keep track of coding decisions, and researchers use memos to track changes in the development of categories, recoding and relabeling the categories as often as necessary. As categories emerge in the process of analysis, the researcher keeps verifying data bits with the categories, verifying as interviews continue, and verifying/confirming with new participants during data collection. If using computers, researchers can easily store records of codes and their meanings at different stages.

❧ WHEN IS IT DONE?

Unseasoned investigators seem to worry a lot about finishing: How will they know when they are done? How will they know when the analysis is good enough? How will they know when they are finished?

You will know. How you know will depend on your goals. For some, the study is finished when they have reached a satisfactory level of explanation, attained their theoretical goals, written up the report, and published it. Some investigators do not consider a study finished until any changes indicated to be necessary by the results are implemented, evaluated, and moved firmly into practice.

Project Histories

Because the goal of all qualitative research is to produce new understanding from data, researchers always require ways of justifying their analyses. Some projects make very limited claims ("I understood what it was like to live in this suburb"), whereas others seek patterns that are easy to justify with rigorous data statements ("This is a problem only for the younger women, and I think I know why"). But even when the goals are limited, the researcher will be helped if he or she can tell the story of how the understanding was discovered and checked out, or why this explanation fits the participants' accounts better than that explanation.

It is always important, in any qualitative method, for the researcher to design the study in such a way that he or she can backtrack to discover and report the history (or development) of the analysis. Most qualitative projects get their claims to validity from the researchers' ability to say how they got there. When the record of a project is only in the researcher's memory or an occasionally kept research diary, it is less impressive than a rigorously maintained, dated, and documented history. But the latter, of course, is also more time-consuming. When you are having a brilliant idea, noting the time of its occurrence may be far from your thoughts.

How do you know what sort of a history is adequate? What you should track will depend on what you are going to rely on. If you don't know what that is yet, you will find it helpful to design simple systems to track some important processes:

※ The gathering of data, how and why samples were selected, the scope of the project determined, how and why new data were sought by theoretical sampling. Documents may have stories in your project that deserve annotation or memos (for example, your understanding of an interview changed when the participant phoned later to elaborate on his story).

※ The genesis of a category or idea. When in the project did it happen? Where did it come from? How did it change your research path? Whether you are working manually or on the computer, this usually means recording a quick memo with any significant ideas, or ones that might prove significant.

※ The story of how an idea develops as you explore more data. (Keep working on that memo—and remember always to date the entries.)

※ The more complex story of how you explore patterns of ideas and links between data and ideas (for instance, when and how you realized, and then checked, that all the older women's experiences were indeed different, and why you settled for this rather than that explanation of the result). This may include notes on how you chased other hunches and showed them wrong or identified cases that didn't fit and developed your theory from studying them. (Write more memos for these insights.)

Audit Trails

The term *audit trail* (or log trail; see Richards, 2005) is used in various ways to refer to evidence that the researcher has kept track of research events and decisions in a way that can be checked by an independent auditor, much as a company's financial books are checked (Guba & Lincoln, 1989). Such a trail is a requirement in some areas and viewed as irrelevant in others; as with many other issues addressed in this book, you need to find out what standards apply in your discipline and method.

What can be audited and what can't? The literature on this subject is still to happen, and the researcher designing an audit process must rely on research principles and common sense. It makes a lot of sense to have someone outside the project review and critique how the researcher arrived at the conclusion. Auditing a project history, either formally or informally, will almost always assist evaluation of the study (by the researcher and by other observers or clients).

On the other hand, it is almost always impossible for an outsider to audit another researcher's coding, because coding, as we discussed in Chapter 6, is a complex process of creation and interpretation of categories and selection of relevant material, and keeping track of that process is impossible. Having an outside auditor code data independently will never justify the coding done. If the coders agree, it must mean that the coding was largely descriptive. If they disagree, we learn something about their different backgrounds and disciplines, but not about the study.

Comparing and Fitting Findings In With the Literature

As a final step in ensuring validity, the researcher must consider how the new findings compare and fit in with the literature. This consideration may take place in the "discussion" section of the research report. The fit, even of new findings, should be easily accounted for and logical. However it is done, the fitting of the findings in with the literature is necessary to ensure a relevant and useful study.

普 DEMONSTRATING RIGOR ON COMPLETION OF THE PROJECT

Reaffirming Legitimacy Following Completion

Projects are reaffirmed following completion through implementation (the findings work), replication (someone else found something similar), and incorporation into a metasynthesis (the findings fit into a conceptual scheme). The pushing of the findings forward by using them as a conceptual scheme for a quantitative study is also important, for it ensures validity of the quantitative study. We should stress, however, that this step can determine new questions (for instance, the distribution of the concepts within the population), but it cannot, as is commonly intended, confirm the findings. Qualitatively derived theory is solid if verified in the process of development and, if conducted well, does not require testing quantitatively.

Triangulating With Subsequent Research

As discussed in Chapter 4, triangulating related projects after the completion of the first project can provide validity for the first project, for if the outcome of the first project had been incorrect, the second project could not build on it. Initial findings would not be verified, and the second project would have to correct the first study. With sequential triangulation, the first study is completed and provides a foundation for the second study.

Reaffirming Through Implementation

Implementing changes suggested by the results of the study provides further testing and often an assurance of the correctness of the results. If, for example, the recommendations of a study suggest a particular intervention for a certain problem, and the problem is resolved following the intervention, this provides further credence for the original qualitative study.

What if the intervention does not work? Such a failure does not provide absolute proof that the results of the initial qualitative study were wrong, but it suggests that this could be a possibility. Certainly, careful investigation should follow.

⋙ SUMMARY

Let's now take stock of the variety of ways in which you can ensure that your project is solid:

1. *Asking the right question:* Regardless of what triggered your interest in a particular topic, after you have conducted a literature search and critiqued what is known about the topic, did you bracket that information? Once you began making data, conducting observations and interviews, were you receptive to revisiting your question and modifying it as necessary?

2. *Ensuring the appropriate design:* Even if you made an informed choice of method by conducting an armchair walkthrough (Chapter 2), once you actually started your study, were you sensitive to the type

of data being collected? Did these data appear to maximize the chances you would have of answering your question?

3. *Making trustworthy data:* Were you reflexive to problems that occurred? Did you work to establish trust with participants and verify data both within and among participants? Are your data dependable? Were you careful to code reliably? To categorize or identify valid themes? As you moved forward in your analysis, were you careful to sample theoretically? To saturate categories? To check and to verify every relationship?

4. *Building solid theory:* As you moved toward developing and refining a theoretical scheme, did you continue to check and verify developing relationships in the data? Did you triangulate with results from other studies? How did they fit? Did you check the literature again? How did your findings support or extend what is already known? How did your findings fit? Are your new findings a possible logical extension of the literature?

5. *Verification or completion:* Once findings are completed, the processes of verification continue. The peer review process in publication (or your committee's scrutiny before you defend your thesis/dissertation) is the first level of approval. Results are also supported (and perhaps modified) when they are implemented or used in subsequent projects.

⧄ RE∫OURCE∫

Berg, B. L. (1998). *Qualitative research methods for the social sciences* (3rd ed.). Boston: Allyn & Bacon.

Bernard, H. R. (1988). *Research methods in cultural anthropology.* Newbury Park, CA: Sage.

Boyatzis, R. E. (1998). *Transforming qualitative information: Thematic analysis and code development.* Thousand Oaks, CA: Sage.

Crabtree, B. F., & Miller, W. L. (Eds.). (1999). *Doing qualitative research* (2nd ed.). Thousand Oaks, CA: Sage.

Creswell, J. W. (1998). *Qualitative inquiry and research design: Choosing among five traditions.* Thousand Oaks, CA: Sage.

Denzin, N. K., & Lincoln, Y. S. (Eds.). (1998). *The landscape of qualitative research: Theories and issues.* Thousand Oaks, CA: Sage.

Denzin, N. K., & Lincoln, Y. S. (Eds.). (2000). *Handbook of qualitative research* (2nd ed.). Thousand Oaks, CA: Sage.

Dey, I. (1995). *Qualitative data analysis: A user-friendly guide for social scientists.* London: Routledge.

Erlandson, D. A., Harris, E. L., Skipper, B. L., & Allen, S. D. (1993). *Doing naturalistic inquiry: A guide to methods.* Newbury Park, CA: Sage.

Flick, U. (1998). *An introduction to qualitative research.* London: Sage.

Guba, E. G., & Lincoln, Y. S. (1989). *Fourth generation evaluation.* Newbury Park, CA: Sage.

Hart, C. (1998). *Doing a literature review: Releasing the social science research imagination.* London: Sage.

Kvale, S. (1989). *Issues of validity in qualitative research.* Lund, Sweden: Cartwell Bratt.

Lincoln, Y. S. (1995). Emerging criteria for quality in qualitative and interpretive research. *Qualitative Inquiry, 1,* 275–289.

Lincoln, Y. S., & Guba, E. G. (1985). *Naturalistic inquiry.* Beverly Hills, CA: Sage.

Marshall, C., & Rossman, G. B. (1999). *Designing qualitative research* (3rd ed.). Thousand Oaks, CA: Sage.

Mason, J. (1996). *Qualitative researching.* London: Sage.

Maxwell, J. A. (1992). Understanding validity in qualitative research. *Harvard Educational Review, 62,* 279–300.

Maxwell, J. A. (1998). Designing a qualitative study. In L. Bickman & D. J. Rog (Eds.), *Handbook of applied social research methods* (pp. 69–100). Thousand Oaks, CA: Sage.

Miller, G., & Dingwall, R. (Eds.). (1997). *Context and method in qualitative research.* London: Sage.

Morse, J. M. (Ed.). (1992b). *Qualitative health research.* Newbury Park, CA: Sage.

Piantanida, M., & Garman, N. B. (1999). *The qualitative dissertation: A guide for students and faculty.* Thousand Oaks, CA: Sage.

Rothe, J. P. (2000). *Undertaking qualitative research: Concepts and cases in injury, health and social life.* Edmonton: University of Alberta Press.

Seale, C. (1999). *Quality of qualitative research.* London: Sage.

Silverman, D. (Ed.). (1997). *Qualitative research: Theory, method and practice.* London: Sage.

Sparkes, A. C. (2001). Qualitative health researchers will agree about validity. *Qualitative Health Research, 11,* 538–552.

Thorne, S. (1997). The art (and science) of critiquing qualitative research. In J. M. Morse (Ed.), *Completing a qualitative project: Details and dialogue* (pp. 117–132). Thousand Oaks, CA: Sage.

Whittemore, R., Chase, S. K., & Mandle, C. L. (2001). Validity in qualitative research. *Qualitative Health Research, 11,* 522–537.

10

Writing It Up

rojects end for a number of reasons: Perhaps funds have been spent, or the period of time allocated to fieldwork has ended; perhaps it is time to graduate, or your family has indicated clearly that it is time for you to find a "real job." Qualitative projects rarely end because all of the researcher's questions have been answered. Nevertheless, whatever the practical reasons for stopping, you need to be able to tell when the project is done. Of course, you have been writing consistently throughout the project. You have been writing field notes and memos; you may have kept a diary and a log trail. If you have been using phenomenology, where the writing is the road to reflection, you will have written many, many drafts. But now these drafts are maturing. The end is in sight, and you must prepare to write drafts, then the final version of a report. This chapter is about writing qualitative research reports. It concludes with consideration of the two writing tasks that often dominate novice researchers' views of their future: writing a thesis and writing for publication.

❧ READY TO WRITE?

In qualitative research, the more you learn, the more you write and the more questions that arise. The end of your project will not be marked by the end of your questioning. The main indicators that you are finishing are that you feel confident about the data and the findings, that you believe they make sense and can be accounted for, and that you can bring your project together, as a whole (Richards, 2005, chap. 9). In short, you should have attained the stage of theorizing (Morse, 1994b) and reached your research goals. You should be able to answer your original research question or account for its not being answered by your project (and you probably have answers to other questions that have arisen during the project).

The work of writing, summarizing, synthesizing, and recording field notes or memos will have resulted in piles of papers, many computer files, transcriptions, photographs, and diagrams. A first task is to take stock of these writings and carefully organize how they will be used (for help with this task, see Richards, 2005, chap. 10). Some of what you have recorded about what you know will already be quite well written, but most will need a lot of work before it can be presented to others. But you know the project is ending because you can make sense of it as a whole. You can *tell* it to a colleague, *write* it in an e-mail, and see a finish line. Now it is possible to *write it up*.

Who Is It For, and Where Will It Appear?

Once the analysis is completed, your first task is to look at the project as a whole and ask yourself: What do I want to say, and to whom? If this project is part of your work toward a graduate degree, the preparation of a dissertation logically comes first, and the professional audience is decided. Usually after the dissertation is done, there will be new questions: How do you publish your work and where?

First, the audiences: Assume there are many (see Richardson, 1990). At this stage, you need to revisit your earlier commitments. Did you promise a newsletter article, a report in a general meeting, or personal communications about the outcomes? You must honor these obligations. (In Chapter 11, we offer some warnings about making early commitments to later reporting.)

Your audience is different from the audiences for which most journalists write. Qualitative researchers most frequently write for other social scientists. They write less often for practitioners or students, and even least frequently for the general public.

Writing Qualitatively

What makes qualitative research more than journalism or a novel, more than a biography or autobiography? A good ethnography will orient the reader and flow like a good novel; a phenomenological or grounded theory article has the "grab" (Glaser, 1978) of a good story, entrancing and trapping the reader.

Whereas good journalism and novels may bring new understanding of human interaction and society, good qualitative research does something different. Throughout this book, we have emphasized the features that separate qualitative research from journalism. The purposiveness of qualitative research and the coherence of questions, data, theory, and outcome give skilled writings of qualitative researchers a different character. As you write, aim for an account of your project that is coherent, strong, and elegant. Good reporting is recognized by the nature and type of description, the depth and focus of the descriptive material, the nature of interpretation, and the level of theoretical development.

These goals should inform everything you write and should result in an account that satisfies the reader. Thick description provides contextual background in which the reader can place the study. It ties together analytic pieces, enabling the researcher to present the components of the analysis logically and sensibly. It is the medium in which the actual descriptions of contexts, phenomena, actions, transitions, procedures, and processes are explained, described, explored, revealed, unraveled, compared or contrasted, and linked to other components.

Before you start to write the final article or report, you must complete the analysis so that you are clear about what you have to write about. Not knowing what to write is the most common cause of writer's block (Becker, 1986). Make an outline, select the segments of your analysis to be covered, and edit data quotations accordingly; prepare any tables, diagrams, or photographs to be used as illustrations, and make notes in your outline where they are to be inserted into your article. The skills you have developed from practicing summarizing and synthesizing are vital at this writing-up phase. When writing, be cognizant of "building a case" (for a good example, see Turner, 1994). Put the scene in context with your preliminary description. Now, systematically and logically, develop an argument. We advise you to outline your argument first, then write within the outline. Remember, like qualitative research, qualitative writing is purposive.

Using Your Data

In qualitative reporting, claims to explanation, interpretation, and theories will stand or fall on the adequacy of the researcher's account of the data and the ways they fit the theory. There is no equivalent of the satisfying statistic that establishes that the researcher is unlikely to have arrived at this conclusion by accident. You must show how you arrived at your conclusion, and why you stand by it.

When Do You Use Quotations?

An early issue is how and when to incorporate quotations from interviews and descriptions from field notes. The goal is to give credence to your claims about the data and provide enough description to bring the situation you are analyzing alive—without having the analysis diluted by long passages that do not address the point being made.

Carefully monitor the role played by quoted evidence. There are two major ways of presenting data to illustrate your analysis. The first is to present your interpretation of findings and then follow with a quotation to illustrate your description. This gives the reader the information to judge whether or not your interpretation represents the data. Alternatively, you may present a quotation followed by your interpretation. This may show your ability to analyze in fine detail, but it could miss the big picture. Pay attention to the reader's need to know where a quotation is from and what it represents, how representative it is, and how much variety there is in the data you are not quoting. As we noted earlier in this book, critical reading of other studies will assist you in setting standards for making and illustrating claims.

Editing Quoted Material

The goal of editing quotations is to maintain and convey the original expression. Recorded interviews may contain portions that are difficult to hear, and despite instructions to type "as is," transcribers may correct grammar, ignore interjections, lose emphasis, or impose punctuation that alters participants' meanings. The researcher must subsequently correct the text to return it to its original form or to remove irrelevant material. And in writing, we select and edit; a report that is a full transcript of all data collection is not a research report. So the question is not *whether* you edit but how to do it well.

Basic rules for editing quoted material are obvious: In editing, you should avoid distortion or misrepresentation of what was said, and you should always indicate where text has been removed (with ellipses) and where you have substituted words (by enclosing them in square brackets). Use italics, capital letters, or boldface type to represent emphasis and to indicate pauses, exclamations, and emotions (such as laughter or crying). Always attempt to contextualize quotations (Where did she say this? Who else was present? What question was she answering? What did she say before this?). A very important skill is to prune any quoted material that is redundant. Passages you find rich and exciting are hard to prune, but they will almost always have more impact if edited. Keep asking yourself if each phrase is necessary. A briefer quotation may be more vivid and powerful and much more relevant to your argument. If you think the participant is implying something, suggest this in your interpretive commentary.

Box 10.1

Criteria for Review Used
by *Qualitative Health Research*

Qualitative Health Research Editorial Review Form

Reviewer: _____

Manuscript no: _____

Please complete the following: If the categories are not applicable, write N/A. This form will be forwarded to the author.

I. Importance of submission: What are the manuscript's strengths? Is it significant? Does it contain new and important information?

_____

_____

II. Theoretical evaluation: Is the manuscript logical? Is it parsimonious? Complete? Useful?

_____

_____

III. Methodological assessment: Inductive approach? Appropriate methods and design? Is the sample appropriate and adequate? Are data saturated? Theoretical analysis? Linked with praxis and/or theory?

_____

_____

(Continued)

(Continued)

IV. Adherence to ethical standards?

_____

_____

V. Manuscript style and format? APA format?

_____

_____

VI. Other comments: _____

Please see additional comments: _____

This section will be retained by the Editor: _____

Recommendations: _____

Accept manuscript as submitted: _____

Tending toward acceptance: _____

Tending toward rejection: _____

Reject: _____

Comments for the Editor:

_____

_____

_____

Using Yourself

In Chapter 5, we discussed the important question of using your own experience as data. As you start writing, this issue will recur. Should you include your personal experience in the reports? As in all questions of qualitative writing, we advise attention to balance. Self-indulgence is to be avoided; keep asking yourself whether your experience is pertinent to your argument, adds evidence or credibility, or merely pads out an unconvincing description. Self-pity is irrelevant; avoid long accounts of how difficult or uncomfortable you found it to gain entry into the field. It's your project and your responsibility. But self-reflection may be very important. We advise you not to dodge responsibility by using the passive voice in your writing. Saying that "the data were analyzed" or "the categories emerged" avoids acknowledgment of your agency.

Brevity and Balance

When you are deeply involved in your data, everything seems to be crucial. How do you decide what is important? A process of distancing is necessary; a reasonable amount of distance will allow you to let go of material that is not pertinent to your argument. This is particularly important when a wonderfully detailed dissertation is being converted to shorter articles. All references may seem relevant, so a first paragraph may contain strings of authors' names interrupting each sentence. Full descriptions of the participants seem deserved, so the reader is given unnecessary detail. The methods section may be too long for an article where the emphasis should be on the results. Attend to these imbalances: You now need to refer to only those works pertinent to this article, describe only those characteristics of participants that matter for this section of the analysis, and briskly outline the methods that contributed to this outcome.

Briefer reports offer the challenge of balancing all these sections with the necessary data. Recall that your categories are masterly compilations of summarizing, synthesizing, and abstracting. The overuse of participants' voices makes the research look like compilations of quotations linked together in a logical way, but with minimal commentary by the researcher. If quotations are overused, the researcher may become silent or backgrounded, and the readers are left to interpret the intent and the significance of the quotations themselves. This is transcription, not qualitative inquiry. A frequent mistake that novice researchers make in writing is to link together quotations from the same or different participants with minimal commentary and then reiterate the obvious in the commentary between the quotations. The resulting description can be shallow, superficial, repetitious, and boring, and, most important, it is unlikely to be analytic.

⟋⟋ RE-REVIJITING METHODOLOGICAL CONGRUENCE

In Chapters 3 and 8, we showed the fit of major qualitative methods with strategies to attain the goals for each method. We stressed that although more than one method may use similar strategies, each method has its unique way of conceptualizing, so that the techniques for the various strategies differ and the final products are distinct.

As you write, check your study for methodological consistency and check your writing for explanations of how this study, in this method, works. If, for instance, you used grounded theory, are you presenting a nice depiction of a process, an integrated theory developed around a core category, elaborated in stages and phases? If you have conducted an ethnography, does it explicate cultural beliefs and practices? Does your phenomenological study provide a description of the essence of a lived experience? Check that your language is consistent with your method. When you are writing for the first time, it is a good idea to find and use as a model an excellent published article that reports on a study that employed the same method you have used. Also, you can ask your critical friends to check that what you have written persuades them.

◊ PROTECTING PARTICIPANTS

You will need to revisit your ethical commitments before you write. Your writing must, of course, adhere to your agreements (made in the informed consent form) to protect the identities of your participants. We discuss some of these issues in Chapter 11, where we address the processes involved in getting your project started.

Ensuring anonymity for participants usually requires much more than merely changing their names. Most frequently, qualitative researchers provide anonymity by removing as many identifiers as possible and reporting demographic characteristics only as group data. In some, but not all, qualitative reports, this does not hinder the presentation of results. Ages, genders, and most characteristics of importance can be reported as means and as ranges for the entire group. When demographic characteristics are presented in detailed tables, participants, shown line by line, are at risk of being recognized. Similarly, attaching the same pseudonym to all quotations from a particular participant makes it possible for someone familiar with the people studied (such as a participant's spouse) to go through the manuscript and trace one person's contributions. In such a case, neither confidentiality nor anonymity is respected.

Protecting participants in this way, however, may severely constrain your reporting. Very often it is important for the reader to know which demographic groups are represented by your account of a response. As you start to write, you may find that protecting your participants exposes your analysis to risk, and that subtler and more creative methods are required. (Lyn Richards once "split" a respondent into two people in the published account of a sensitive project!)

Quite often, researchers replace the names of towns or institutions with fictitious names in their writings. Sometimes, however, authorities at research sites agree or request to have their sites named in final reports and in publications; if this is the case in your research, be sure to obtain this agreement or request in writing. Carefully consider the advantages and disadvantages before you agree to name an institution or participants— doing so may place you in an awkward position. If the institution asks to be named, what do you do with any findings that may reflect negatively on that institution? Omitting such findings may result in an invalid report, and including them may leave you open to criticism or, in the worst case, a lawsuit. This is a dilemma that you must consider carefully. On the other hand, Davis (1991) argues that removing too much contextual information (including important identifiers) may render a case study useless and invalid. Clearly, a balance must be attained, but the researcher's primary obligation remains the protection of participants and research sites.

You may be asked by the organization in which you have collected data to present administrators with a draft report for "approval." When negotiating for entry, do not agree to give the organization the opportunity to critique your report, for doing so may result in your conclusions being undermined and invalidated. If you have done your study with care, are confident about your results, and have "cleaned" your findings so that the organization and the participants cannot be identified, you should be able to write results that are insightful and justifiable.

〴〴 EVALUATE YOUR WRITING

Once you have drafted a chapter, an article, or a report, evaluate it yourself (and perhaps ask your colleagues to review it) to ensure that it is as solid as possible before submitting it. Have your article professionally edited. Many of the works listed in the reference section at the end of this book include discussions of criteria for evaluating qualitative research (detailed advice and checklists for evaluating your writing are in Richards, 2005, chap. 10). Evaluation criteria, of course, will vary with the method, but all qualitative methods share two criteria for excellence.

The most obvious criterion is that a report should offer knowledge that is new. This does not, of course, mean the report will be entirely new. Researchers who avoid established labels for theory can end up merely relabeling old theory. In the last stage of your analysis, place your findings into the context of the literature, involving the previous theory as you identify unique contributions of your work. But be clear about what you offer that is new, and clearly offer it.

The second criterion involves the nature of the analysis, the coherence of the theory, and the elegance of presentation. If the author's intention is to produce theory, then the criteria for good theory should be applied. Is it clear, logical, and well argued? Is the theory parsimonious—tight and elegant? Is the theory coherent and internally consistent (see Morse, 1997)? Evaluation of the research methods and the contribution of the research as a whole are also in order (Kuzel & Engel, 2001; Thorne, 1997).

⧬ POLISHING

Once you have written your "final" draft, leave the manuscript alone for a few days to mature, then take it out and read it aloud. It will probably need heavy editing and rewriting. You will be able to identify skimpy pieces or parts that are simply too wordy. If you are writing an article, look again at the criteria for review from the journal you are submitting to, and, using those criteria, review the article yourself. When the article is as perfect as you can make it, give it to three colleagues to critique, along with the journal's criteria. Then revise again, incorporating their suggestions. Finally, have your article professionally edited, make the required number of copies, and submit it.

A final word of advice: We argued early in this book that qualitative research is a craft. By now you will have gathered that writing qualitatively is an art. Your reading of published qualitative work will have shown the high proportion of studies that fail to convince because they are presented inadequately. This is because writing up qualitative research, like all art, requires vision, focus, and hard work—if done well, it is extraordinarily satisfying for both the artist and the audience.

⧬ USING YOUR SOFTWARE FOR WRITING

Your writing will of course be done on a computer. Word processors have remade the tasks of composing, editing, and revision.

Approaches

Your word processor will easily receive the data and reflections from your project if it has been created and managed in qualitative software. At any stage when you are writing, you should use the ability to make reports of your data or the passages coded or discovered in your analysis.

Advances

Throughout this book we have encouraged the recording of your project and your steps in analysis in a "trail." If appropriate, this can include the recording of checks you have made on coder consistency and sample diversity. This "log" can be very efficiently kept in your software, with links to the relevant data or categories described. (For detailed assistance in setting up a "log trail" in software, see Richards, 2005, chaps. 1–3.)

Your software's search tools have many tasks in this stage (Richards, 2005, chap. 8). Use them for locating relevant data or quotations, checking the patterns you are discerning and how adequate a coverage you have of critical issues, or finding exceptions to a generalization. Don't underestimate the uses of text search to check whether a "dominant" theme really is dominant!

As you start the final reporting task, make good use of your software's data management and search tools to conduct a stocktaking of your memos and log trail writings so that you take account of all the insights or concerns recorded during analysis (Richards, 2005, chap. 10).

Alerts

Be aware that with your data available online, it is much easier to use too much quoted material. Patchworks of quotation do not substitute for your good writing of your analysis.

Be very careful of the language you use as you interpret and discuss the results of software-supported search processes (Richards, 2005, chap. 8). Software searches are often very powerful and can certainly offer discoveries and checks not possible by manual methods. But they should always be reported for what they are—mechanical searches based on the text *you* provide in your project or the coding *you* do. (You are the weak link!)

Schedule your time to include using your software well to provide material and check it out. Your analysis will be much stronger for this work. Don't be tempted to dodge it and merely state your hunch and illustrate it with a juicy quotation—to do so will save time now, but it will cost time when you are required to resubmit your report!

And finally, expect that it will be harder to finish when you know there is always another search or report you could do. Plan the processes needed, schedule them sensibly, evaluate, edit, strengthen evidence, check out patterns and explanations claimed—and finish!

※ WRITING YOUR THESIS OR DISSERTATION

A student's thesis or dissertation is often the first major book-length manuscript, first piece of completed research. If that is not anxiety provoking enough, a lot hinges on the student's doing a good job—the manuscript will be scrutinized by several faculty members and an outside examiner, and if the research is not up to expectations, an entire career is at stake! Fortunately, the final examination is not something that happens after several years of working alone—it is usually a "journey" that is shared by your supervisor. Graduate school is (or should be) a wonderfully mentored situation. During the conduct of your research, guidance is available and this will continue as you write—so consult if you are not certain about any aspect. Furthermore, in your department and library, there should be many theses/dissertations available to use as models to help you see what the final product should look like. Use also the new resources provided by the Internet for access to help and advice via professional discussion groups and forums.

Another consoling feature is that some of your dissertation may have been already written in the process of doing research: Your proposal may contain a literature review; it may need updating, but this preliminary work may at least form the basis of the literature review for your thesis. Similarly, the methods section for your proposal may be used as the basis of the methods chapter for your dissertation, but remember to not merely change the tense from the future tense of the proposal to the past or present tense of your thesis/dissertation, but also to incorporate what you have learned as the project unfolded. Qualitative methods chapters (unlike those for survey projects) can never be written in advance of the project, since the method requires that the project design is flexible and data driven. If you have been carefully recording references since your proposal, the *Bibliography* should be in the proper format and relatively complete. These pieces, no matter how small, are an excellent start for the final document.

This section is a walkthrough of the process of actually *writing the dissertation*. Note that dissertation requirements vary greatly by country, by institution, and by discipline. In the rest of this section, Morse draws on her experiences.

A dissertation is usually organized in a conventional outline, and the form and format, structure, and even the style may be dictated by regulations from your graduate school. Obtain these guidelines before you commence. The overall outline may look like the one in Figure 10.1. You could safely use this to frame the details of your own study.

Beginning to Write

While it is tempting and obvious to start at the beginning, smart writers leave Chapter 1 until later in the project. In qualitative research, writing is a way of analyzing, and the project may continue to take shape as you write. Hence it is important to be writing—in memos from early in the analysis process, and later in drafts of your chapters—-and to take very thorough stock of all your writing as you near the end (Richards, 2005, chap. 10).

It is difficult to introduce something that is not written yet, and not knowing exactly *what* to write is one of the most common causes of writer's block. So start writing your dissertation with the *Results* chapter.

The easiest chapter to write is the *Methods* chapter, because it is very structured and contains contents that are relatively concrete; next write the *Literature Review* chapter, followed by the *Discussion* chapter, linking the discussion of the findings with the literature (Chapter 2) being especially clear about what your study has contributed to the literature. While writing, constantly keep the bibliography up-to-date and accurate. Then, once Chapters 2 to 5 have been written, you will find that writing Chapter 1, the introduction, will flow easily and naturally. Write the final summary (of 3 to 4 pages), and then summarize that summary for the Abstract.

Great! But we are not finished yet! Once you have completed the *draft*, edit it yourself. Read it carefully and slowly, front to back, making comments as you go. Now edit your draft—and be particularly critical of your *Results* chapter(s). Make your corrections only when you are satisfied with your draft, and ask your colleagues and friends to review it. Have it professionally edited, and only when it is possible, give it to your committee to review; then prepare for your final examination. Good luck!

1. Descriptive title

2. Front signature/approval pages

3. Abstract

4. Table of contents

5. Preface/foreword

6. Chapter one: Introduction
   a. Problem statement

7. Chapter two: Literature Review
   a. Introduction to chapter
   b. Components of the topic reviewed
   c. Summary/framework/research question

8. Chapter three: Methods
   a. Introduction to methods chapter
   b. Description and justification of method
   c. Setting and sample
   d. Data Collection
   e. Data analysis

9. Chapter four: Results

Here you may systematically describe the results, attending to the pointers previously given about *writing qualitatively*. You may need to break this content into two chapters if it gets too long. Sometimes researchers present the descriptive results in one chapter, and then the content pertaining to building their qualitative theory is placed in the second chapter, Results.

10. Chapter five: Discussion

A discussion chapter is often required; here you would consider how your study fits in with and contributes to the literature. Include a discussion of the strengths and the weaknesses of the study; include a discussion of the methods used. How could your study have been improved? What should the next study be?
Conclude with a comprehensive summary of the entire dissertation, usually brief, about four pages in length.

11. Appendices

If you used any questionnaires or other research guides, they usually are placed here, along with the concept forms, letters of approval, and permissions to conduct the research.

Figure 10.1   Sample Outline for a Thesis

Readers have requested a source of published qualitative dissertations that may be used as models for their students. The *International Institute for Qualitative Methodology* (IIQM) conducts an annual international dissertation award, and the winning dissertations are published by the *Qualitative Institute Press* as monographs. These are available through the IIQM website: http://www.uofaweb.ualberta.ca/iiqm/

〉〉 WRITING AN ARTICLE FOR PUBLICATION

Having written a great dissertation, and being convinced that all of the content is important, it is tempting to write an article by simply cutting sentences and even paragraphs and placing them directly into your article. This seldom works, for the style of writing used in the dissertation is very different from the style used in shorter, more succinct articles.

When writing an article from a dissertation, the first thing you must do is to decide on the content. For instance, will the article be a summary of the entire dissertation, or some portion of it? This is not an easy decision, for the most obvious way to partition articles is to sever (note the violent word) the dissertation into concepts, publishing an article from each segment as a meaningful unit. But this means, after carefully arguing in your dissertation for the holistic qualitative perspective, that you are now fracturing your theory into small chunks. Alternatively, you may publish the entire theory intact, but due to the page limitation, it will be relatively superficial. Or you may consider publishing the whole study as a monograph. Another common way of dividing a project is by publishing the literature review as a discussion and critique on the topic, and then the results as a project report. Sometimes the project will also support an article on the implications for those practicing, such as clinicians or teachers. These decisions should be made early, as editorial rules and copyright regulations often prohibit duplicate publication.

Next, select your audience. Qualitative researchers most frequently write for other social scientists. Less often do they write for practitioners or students, and even less frequently for the general public. Because selected audiences read certain journals and have distinctive styles, select the journal in which you wish your article to appear. Obtain the appropriate author guidelines from the journal, and read back issues of the journal to familiarize yourself with its style (especially its referencing style) and content. If you are writing for practitioners, you will note that the emphasis of such articles is on the application or implications of the results, rather than on the methods and scientific aspects of the study. If you choose to write for a wider public, seek advisors, editors, and readers who can help you translate your academic research project for readers with little concern or interest in the study's methodological origins or specialized concepts or theories.

Beginning to Write

Writing an article requires following the general procedures for writing a thesis/dissertation. First, prepare a detailed outline; prepare and edit all the quotations that you are planning to use and enter them into the outline. Similarly, prepare any tables or figures that you plan to use, and temporarily place them in the outline. (Publishers' formats in manuscripts usually require that the author indicate in the text where the figure or table will be placed and insert the actual table or figure at the end of the document.) Thus we see in manuscripts instructions such as: "Insert Table 1 about here."

The order for easy writing is similar to that for dissertation writing: Start writing in the *Results* section, followed by the *Methods, Literature Review, Discussion, Introduction,* and *Abstract* sections. Attend to the proportions of text per section as you write. For a 15-page article, the *Results* should take up the bulk of the manuscript (at least 7 pages); the *Methods,* 1½ pages; the *Literature Review* and the *Discussion,* 2 pages each; and the *Introduction,* 1½ pages. Cite only references that *deserve to be cited,* and keep them to a minimum. Some journals limit the number of citations (for instance, *Advances in Nursing Science* limits citations to 64, sometimes requiring authors to make difficult choices for providing proper attribution).

Once the writing is finished, put the article aside for a few days, so that when you return to edit it, you will see it with fresh eyes. When you do return to it, read it out loud. This will enable you to hear all its faults, and then edit carefully.

Most journals have a website with *criteria for review.* These criteria are the dimensions by which the reviewers will evaluate your article. Print out these criteria, and review your own article, making changes accordingly.

What are these review criteria? For *Qualitative Health Research,* the criteria are shown in Box 10.2. Note that these criteria resemble the *standards* for qualitative inquiry that are discussed in the literature. Following these criteria will ensure that you have all of the components needed for the publication of an outstanding article, and that each of these components is as solid as possible. After these revisions, one more review is necessary: Give your article and the review form to three colleagues whose judgment you trust, and ask them to review your article. Again, make any necessary changes.

Box 10.2

Something to Do

When reviewing the literature, select several studies that you find to be interesting and well written, and critically analyze how the authors present these studies.

- How do the authors provide their evidence?

- When is the evidence persuasive?

- How do the authors capture their readers' interest?

- Do the authors build their cases convincingly?

- How do they present their interpretations?

- How do they link their findings to the literature?

- What does not satisfy, and how could the authors have rectified these problems?

You can learn from the experts this way.

While the review and revising seem to be time-consuming, it is a process that will save you considerable time in the long run, and it may make the difference between having your article accepted or rejected. It means that your article will move into press as quickly as possible and not be unnecessarily delayed.

Once it has been submitted to a journal, what happens? If you do not receive notification after a few weeks, contact the editor. Do not sit waiting to hear about a manuscript that was lost in the mail and not received by the journal!

After those weeks of waiting, you will be notified that your article (a) needs further revision, (b) has been rejected (you know what to do—revise and submit it to another journal), or (c) has been accepted.

If the article has been accepted, congratulations. Your notification will include a manuscript number that must be used on all correspondence with the editor.

The journal will request the copyright for your article, and this may require your completing a form. Yes, the publishing company owns the copyright, not usually the author. Next, after several weeks or months, you will suddenly receive the proofs. Your article will have been edited by a technical editor, and as this editing process sometimes changes the meaning of sentences, check every word carefully. You will have been sent a guide of proofreaders' marks, instructing you how to mark any necessary changes on the manuscript. Do this as quickly as possible, as the turnaround time is often very short, 72 hours or less. Again there is a wait, until suddenly, one day you will receive several copies of the journal in the mail containing your published article. This is one of the most exciting moments—read your article and put one journal away carefully, give one to your Dean, and a third one to your mother!

\\\ AFTER PUBLICATION, THEN WHAT?

Once your study has been published, it may take on a life of its own over which you have little control. It is exciting to see your work cited by others, or to hear your writing quoted. But often we wish that our findings would be implemented and used by others.

We may dream that our research will make a difference, but that rarely happens from a single project; but if that does happen, changes usually occur slowly.

Qualitative research findings may be adopted by others and used in a number of ways. Of course, how your study is used, and what impact it has, depends on the type and relevance of the results themselves. The most common ways that qualitative inquiry is used are alone or in concert with other findings.

Findings That Are Used Alone

Well-designed qualitative projects create results that *are* useful, transferable, and generalizable. *How* they may be used, of course, often depends on others. Once the findings are published, as noted above, the original investigator has little control. Possible uses for qualitative findings include the following:

Use the Theory as a Framework for Practice

The theory created in your study may be adopted into teaching, counseling, or professional practice and used to organize the way others teach, counsel, or practice. It may be used to explain outcomes that occur in practice or to provide understanding about what is going on. In evaluation research, it may, for example, be used to indicate what changes should be made in an organization to improve the workplace. Overall, how the findings are transferred and utilized will depend on their presentation during dissemination, how clearly they are understood, and how easily they can be put into practice.

Bring the Implicit and the Informal to the Fore

Qualitative findings often document the unknown or the unacknowledged parts of the context, or the previously unrecognized actions of participants, making overt what was previously covert. For example, they may show the taken-for-granted assumptions about power in a relationship and allow participants to confront and discuss them. Thus, by bringing such behind-the-scenes actions forward, new understanding about a setting, a process, or a procedure may be recognized, and thus incorporated into the understanding of others.

Delimit Scope or Boundaries of Problems or Concepts

Because qualitative researchers often address problems that have been previously overlooked, the research may allow problems that have seemed insurmountable, or not suitable for investigation, to become accessible. Qualitative research, therefore, may open up new areas of investigation. In the process of qualitative description, these problems become deconstructed. As problems are partitioned, concepts are identified, and the initial description thus enables others to attempt further inquiry.

Describe the Problem and Aid in the Identification of the Solution

Often the initial "thick description" from qualitative inquiry may be used to identify possible solutions or interventions for the original problems. If you are studying a certain concept, for instance, the concept of hope, your results will focus on the participants' meaning of the concept and their behaviors working toward whatever they are hoping to attain. You will probably not have, for instance, much data on caregivers, how they fostered, modified, or assessed hope in their patients, because the caregivers may not have been a part of your original study. But you may extend your findings theoretically to create an assessment guide. An observation about hope may be *"recognizing the threat."* This may be transformed into behavioral indices: *"Did the impact of the event sink in?"* referring to observable behaviors, such as "Reiteration and repetition in speech; connecting with others to talk about the event; and appearing stressed and overwhelmed" (Penrod & Morse, 1997). Thus, an assessment guide may be created from the qualitative results.

Provide an Evaluation of Nonmeasurable Interventions

In many research areas, traditional methods have insisted that the only adequate way to demonstrate efficacy is to conduct a randomized controlled experimental trial that will give a definitive answer. However, qualitative description, when applied to change models, program implementation, or other dynamic situation, often provides information and results that offer a quite different understanding from statistical analysis and may enable assessment of significant serendipitous outcomes.

The Cumulative Effect of Research Results

The most common outcome of qualitative study is that the findings fit (agree/support) with the published findings of others. This is the way that science progresses, and other investigators will add your study to their literature reviews when preparing the background literature for their own studies. If your study, on the other hand, contradicts others, this may make yours particularly interesting and something to be investigated further. This may be done within your own program of research or by others. One of the most exciting results is that the next study—or series of studies—then appears. Research has a habit, nasty or otherwise, of trapping you in a career trajectory!

※ SUMMARY

Qualitative research involves constant writing—of field notes, memos, interpretive definitions, and team notes. We have urged you to see writing as ongoing from the start of a project and "writing up" as something that matures during a project. Use all those materials well as you reach the end of the project. Once you are able to theorize and are confident that you understand what is going on, the writing up of the project commences. First, identify who you are writing for and what you want to write. Plan an outline; prepare data, logs, and diagrams; and insert them into the outline—then write! As you write, consider the audiences and purposes, and possible future outcomes, of your work and the ways you are now reporting it.

※ RESOURCES

Theory

Morse, J. M. (1997). Considering theory derived from qualitative research. In J. M. Morse (Ed.), *Completing a qualitative project: Details and dialogue* (pp. 163–188). Thousand Oaks, CA: Sage.

Strauss, A. L. (1995). Notes on the nature and development of general theories. *Qualitative Inquiry, 1*, 7–18.

Evaluating Qualitative Research

Altheide, D. L., & Johnson, J. M. (1994). Criteria for assessing interpretive validity in qualitative research. In N. K. Denzin & Y. S. Lincoln (Eds.), *Handbook of qualitative research* (pp. 485–499). Thousand Oaks, CA: Sage.

Leininger, M. (1994). Evaluation criteria and critique of qualitative research studies. In J. M. Morse (Ed.), *Critical issues in qualitative research methods* (pp. 95–115). Thousand Oaks, CA: Sage.

Muecke, M. (1994). On the evaluation of ethnographies. In J. M. Morse (Ed.), *Critical issues in qualitative research methods* (pp. 187–209). Thousand Oaks, CA: Sage.

Thorne, S. (1997). The art (and science) of critiquing qualitative research. In J. M. Morse (Ed.), *Completing a qualitative project: Details and dialogue* (pp. 117–132). Thousand Oaks, CA: Sage.

Reliability and Validity

Kvale, S. (1995). The social construction of validity. *Qualitative Inquiry, 1,* 19–40.

Sandelowski, M. (1993). Rigor or rigor mortis: The problem of rigor in qualitative inquiry. *Advances in Nursing Science, 16*(2), 1–8.

On Writing

Boyle, J. (1997). Writing it up: Dissecting the dissertation. In J. M. Morse (Ed.), *Completing a qualitative project: Details and dialogue* (pp. 9–37). Thousand Oaks, CA: Sage.

Penrod, J., & Morse, J. M. (1997). Strategies for assessing and fostering hope: The *Hope Assessment Guide. Oncology Nurses Forum, 24*(6), 1055–1063.

Richards, L. (2005). *Handling qualitative data: A practical guide.* London: Sage.
 See especially Chapter 10, Telling It.

Richardson, L. (1990). *Writing strategies: Reaching diverse audiences.* Newbury Park, CA: Sage.

Richardson, L. (1994). Writing: A method of inquiry. In N. K. Denzin & Y. S. Lincoln (Eds.), *Handbook of qualitative research* (pp. 516–529). Thousand Oaks, CA: Sage.

Part IV

BEGINNING YOUR PROJECT

11

Groundwork for Beginning Your Project

Throughout this book, we have aimed to offer an overview of what goes into doing qualitative research. Now that you have a feel for the process, let's go back to the beginning. To start a project, you must first identify a topic of interest, go to the library and learn all about the topic, and, in light of what is already known, refine your research question. An armchair walkthrough will help you think through your question using several methods. This enables you at this early stage to be as certain as possible that you are making informed choices and to foresee any possible problems in selecting your methods and your research design. Most important, it gives you an insight into the type of information you may find in your results.

Once you have decided on a topic, chosen a research question, and put it in the context of the pertinent literature, the usual next step is to work out a research design and prepare a proposal describing your research. Depending on your research setting, this may be a formal proposal to be approved by a supervisory or ethics committee. But even if you are not required to present a proposal, we urge you to prepare one as a step toward clarifying your thinking and your purpose.

※ WRITING YOUR PROPOSAL

What is a proposal? It is a document in which you outline, to the best of your ability, what you plan to study, why, and how you will do it. It serves as a guide to inform your dissertation committee; it serves as something to be evaluated by ethics review committees to determine if you will possibly harm participants; it serves to inform interested parties in the institution or community where you will be conducting the research; and it serves as the baseline for the start of your audit trail. Several authors have written guides on preparing proposals for qualitative researchers (see Boyd & Munhall, 2001; Cheek, 2000; Morse & Field, 1995).

A quantitative research proposal is a rigidly formatted document that describes in detail ordered procedures to be adhered to during the research project and promises certain outcomes. By contrast, a qualitative research proposal serves a different purpose. As the purpose of the qualitative project is discovery—to find out what is going on—the researcher cannot predict or promise a certain outcome from the research (Boyd & Munhall, 2001; Morse & Field, 1995). The proposal should make an argument for doing the study and convince the reader that the topic selected is significant and worthy of inquiry. It must describe the method to be used, the research setting and participants, what the researcher will do to gather and handle data, and the intended data analysis strategies.

Using the Literature Review

The literature review usually comes first in a proposal, offering an overview of what is known and what is indicated by previous research. It locates the proposed project in the current body of knowledge. The literature review should seek to reveal (rather than to conceal) gaps in knowledge and to show areas that are weak and lacking or results that are suspect and perhaps built on assumptions that may be queried. Thus the literature review leads the reader toward the research question, so that by the end, the question reads as an imperative that must be urgently investigated. If yours doesn't, you need to rethink your question and your rationale for asking it, so your project will be useful and contribute new knowledge.

Although the outcome of qualitative research cannot be predicted, the significance of the project itself may be couched in terms of a *theoretical context* derived from the literature (Morse & Field, 1995). The theoretical context is larger than the proposed project, but it places the study in the context of the topic. For instance, if the proposed project is about social support that occurs between patients, literature on the significance and efficacy of peer support in other contexts might locate the study in works on the value of patient-patient support (or factors that impede peer support). Without peer support, we cannot justify many practices, such as support groups. We could even argue that the study will have implications for facilitating patient-patient contact or for hospital design. In essence, the theoretical context is a persuasive argument that extends beyond the proposed research problem and shows the possible ramifications of the study and the way it fits into the greater scheme of things.

Writing the Methods Section

From your armchair walkthrough, you will have a reasonable idea of the method that will best address your question, the setting in which you want to conduct your study, and the types of participants and procedures you need for making and handling data. In your proposal, you must describe these as clearly and with as much detail as is necessary for a reader to assess their feasibility and adequacy. Justify your choice of setting and your selection of method. Explain that when sampling, qualitative researchers maximize access to the phenomenon they are studying and select cases in which it is most evident. Obtaining such a "pure" sample is consistent with the principles of science. Build into your plan the expectation that once you have gained an understanding of the phenomenon, you will employ *theoretical sampling* (Glaser, 1978) to seek out particular variations by selecting participants according to the theoretical needs of the study.

Because there is no formula for determining sample size in qualitative inquiry, you are unlikely to be able to cite exact numbers for the size of the sample. Instead, explain that the number of participants recruited will be determined by the quality of the participants' experiences, the ability of the participants to reflect on and report their experiences, and the requirement for further theoretical sampling. Explain how data collection will cease once saturation is reached and what the indicators of saturation may be.

Next, show how you will handle and analyze data. You must present data-handling and analytic procedures in sufficient detail to give reviewers confidence that you will be able to do justice to these data. Indicate your understanding of the sorts of coding and analysis your method requires. If you intend to use a computer program, specify it and justify your choice. (Never merely assert that the "data will be analyzed by" a particular program—software, remember, does not analyze data.) Depending on the detail required, you might illustrate procedures by, for instance, inserting a sample of text with actual coding. Demonstrate how your analysis will proceed with building categories, theme-ing, or using memos and annotations.

Estimating Time (and Related Resources)

Any qualitative project has many stages, and we strongly advise that you consider all of these necessary stages from the start. Possibly the most common mistake of novices is to misjudge the time that research will take. Serious crises are created when a researcher plans the use of available time or money by calculating time in terms of data-making events (such as the average length of an intended interview multiplied by the expected number of interviews to be conducted). A more realistic calculation would take into account the usually substantial time needed for each of the many stages before and after interviews (or other data-making events) and also the amount of research work generated by each event. In Figure 11.1, we sketch the demands on your time during a project. Using this formula, you can more clearly see the next year(s) of research time and more realistically view, scope, and rescope a project. You may wish to start by taking the time to revise the scope of your project right now as you face the time that preparing your original proposal would realistically require.

Developing a Budget

If you are applying for a research grant, you are required to cost a proposal. If you are not seeking a grant, costing is possibly an even more important task. Your salary stipend or ability to support yourself through the project will be contingent on your estimation of overall time (above), but other costs must also be estimated. Pay careful attention to the following:

◎ *Personnel:* For the salaries of research assistants, estimate hourly rate × duration of employment. Remember to add library time, time for scheduling appointments, travel time, time for listening and checking transcripts, time for involvement in analysis and preparation for subsequent interviews or the next stage, and time for copying, filing, and other clerical work.

◎ *Transcribing:* If tapes are to be transcribed, this will be a major cost. (Estimate the number of interviews × 4 hours × 60-minute tape × hourly salary.)

Time for research design, critique and discussion of the design, and later for redesign.

+

Time for conceptualization, getting into your topic, literature review, and critique.

+

Time for preparing, making a reconnaissance, entering the field + time lost by making mistakes, being adopted by the wrong people, misinterpreting early signs, and putting your foot in it.

+

Time for choosing, trialing, and setting up a data management system, including time to get good at it and at underlying skills in the library or on the computer.

+

Time for ongoing data management (backing up files, setting up a computer project or labeling boxes, archiving records, descriptive coding).

+

Data creation and interpretation time. For each event (e.g., an interview), allow time it takes + time for preparing, practicing, learning, rehearsing, and losing your nerve + time making the appointment, finding she isn't home, and remaking it + time for transcribing audio or video files (at least 5 times the time the event took) + time for checking transcripts, researcher time for reading, interpreting, and coding data.

+

More data collection and interpretation time for planning theoretical (data-driven) sampling + time for each of these new events and the processing accompanying them.

+

Time for ongoing coding, exploring, annotating, categorizing, and, above all, thinking, trialing, and establishing explanations, despairing, exalting, talking, consulting, revising, and revisiting.

+

Time, from the earliest stages, for analytic work: writing, annotating, memo writing, rewriting, editing, getting it right, discussing, and refining.

+

Time for reporting, to participants or professional audiences, in as many ways and contexts as the project deserves.

+

Time for the party with your team.

Figure 11.1   Estimating Time for Qualitative Research

◊◊ *Equipment and training:* Present the makes, model numbers, and prices of all tape recorders, transcribing machines, computing equipment, and the like. Allow for time and costs of training in interview techniques, computer methods, coding, and so forth.

◊◊ *Payment for participants:* If you plan to pay participants, add costs for focus groups or interviews (usually by time committed), plus reimbursement of expenses for parking, transportation fares, baby-sitting fees, and so on. Allow for expenses for refreshments.

◊◊ *Supplies and travel:* Think ahead to estimate expenses for software, paper, pens, computer disks, audio- and videotapes, printing cartridges, film and developing charges, and copying. Also budget for travel for yourself and any research assistants.

◊◊ *Publication costs:* At this stage, publishing may seem a distant dream, but it will eventually entail costs. Plan on expenses for editing, graphics, duplication of reports, mailing, and courier services.

A Note on Dealing With Available Data

Perhaps the data you need are offered ready-made. A preemptive research design, planned before your involvement, or data available for secondary analysis are always a challenge, and sometimes a major risk. This is not to argue that you should refuse to analyze the data already available. Such a situation occurs in many projects. Historians live with it, routinely using documents, diaries, letters, or other data that predate their projects as a way to explore the past. Nor do we urge you to refuse when you are offered data—by an activist group, for example. Some of the great qualitative studies happened that way, with researchers responding to available data. If this is your situation, our best advice is that you take your research design even more seriously than you would if the project were proceeding with data-making processes under your control.

However, available data are highly unlikely to be the data you want or a perfect fit for your question. Your proposal must contain a budget and time for working with the data or possibly gathering additional data. If you want to use a secondary data set, sometimes you will need to think your way into the project and the data set, assess very sharply the adequacy of the data available to you by these means, and make claims in the context of the data's limitations.

※ ENſURING ETHICAL REſEARCH

Qualitative inquiry brings with it special issues pertaining to participant consent and maintenance of participant anonymity. In almost all research settings, formal ethics review committees examine researchers' proposals for inherent risks to participants, specifying the level of consents and permissions required from participants and community leaders and the level of anonymity that must be provided to the participants. Before beginning their projects, researchers must obtain permission from their employers (or, in the case of students, from their universities). The institutions in which research will be conducted may also have their own rules regarding access and other issues (as do most schools, hospitals, prisons, and government departments). Researchers often complain about the restrictions placed on them by ethics review committees, but we urge you to give thoughtful consideration to the need for such restrictions. Working outside such a context, you would have the much more difficult challenge of ensuring ethical practice without the assistance of a committee.

Your research institution (and host institution) will have its own ethics procedures and forms; you must use them, but you should also make sure that they adequately cover the requirements of your project. In addition, you must meet the federal requirements for your country; information about these is generally available on the Internet (for the United States, see National Institutes of Health, 2001; for Canada, see CIHR/NSERC/ CRSNG/SSHRC, 2001). A sample of a participant consent form appears at the end of this chapter.

The Challenge of Anonymity

We warned in Chapter 10 how easy it is to breach anonymity by using blocks of quotation and dialogue verbatim in the final report. More extreme risks accompany the use of photographs or videos. Participant consent forms must state the mode of recording and the planned uses of records. Because participants may be recognized in photographs and videos, thus violating promised anonymity, you will need to obtain releases (with separate signature lines) if you are taking photographs or videotaping (see the sample consent form addendum at the end of this chapter).

If you have promised your participants anonymity, you will need to check all written material from the start to ensure that persons and places are not recognizable. This is no trivial task, and it is rarely achieved merely through the changing of names. You must make certain that no participant's anonymity is violated indirectly through the linking of his or her demographic characteristics, such as age, gender, marital status, occupation, disease, and even pseudonym.

Unless the participants actively insist on being identified, it is standard practice to change proper names of people, suburbs or villages, and institutions. Even if participants seek identification, you should not identify them unless they have signed releases allowing their names to be used. (And in such situations, you should still be acutely aware of the effects of publishing your research.)

Important to the issue of anonymity is the researcher's responsibility for concealing the names of the institutions or locations in which the research was conducted. We discussed these issues in Chapter 10.

Permissions

Since the 1980s, social science research has been carefully monitored to protect human subjects from any risks (including invasion of privacy). A researcher cannot undertake a project without first getting a series of permissions and having the research proposal reviewed by university committees and the agencies and communities involved.

First and foremost, you must obtain permission from your employer, for the institution that employs you is considered to be primarily responsible for your actions. Even students (who are not involved in a direct employer-employee relationship) must obtain permission through their universities' ethics review procedures before they can negotiate sites for their research. So your first task, after completing your proposal, is to find out—and follow—your own institution's guidelines. A university's permission is usually reviewed and renewed annually and requires that any incidents and complaints be immediately reported.

The second step is to get support for and permission to conduct your research at the site you have selected. Permissions here are usually obtained on several levels. First, you must obtain approval from the top administrative level, and this often entails a second ethics review. A hospital, for example, will review your proposal to see what you will be asking of staff and patients. Administrators will be concerned about how much staff time is involved in your project (for the institution, staff time is a cost) and what will be asked of patients. They will consider your proposal in light of other ongoing research, for they are also concerned about research burden—the amount of research currently being conducted in the institution and its impact on patients and patient care.

International research is a special case. If your project involves your traveling out of your own country to collect data, you will need special permits and research visas from the host country. These may take some time to obtain, so you should allow ample time for this process in your proposal.

The final level of approval must be obtained from the actual research setting, such as the unit or the classroom where the study will be conducted. Please be aware that getting approval at this level does not mean that the staff will automatically support your research. Once in the setting, you must obtain individual consents and permissions; you must also fit in and win the trust, cooperation, and support of the staff.

Participant Assent and Consent

Covert research (research that is carried out without the knowledge of participants) is rarely approved by ethics review committees. If it is necessary for research to be conducted without the participants' knowledge, extensive debriefing procedures must be established and consents obtained after the collection of data, with participants given the option of having data referring to them destroyed if they do not wish to participate in the study.

Participants' rights include the following: (a) the right to be fully informed about the study's purpose and about the involvement and time required for participation, (b) the right to confidentiality and anonymity, (c) the right to ask any questions of the investigator, (d) the right to refuse to participate without any negative ramifications, (e) the right to refuse to answer any questions, and (f) the right to withdraw from the study at any time. Participants also have the right to know what to expect during the research process, what information is being obtained about them, who will have access to that information, and what it will be used for.

Researchers are usually required to obtain formal written consent from participants prior to the commencement of the data-gathering period. To ensure anonymity, the consent forms are not linked to the data in any way. Researchers commonly keep the forms for a period of 7 years after the study is completed.

If a participant is a minor, usually defined as being under the age of 18 years, special conditions apply. If a child is not old enough to understand the concept of research or the required protocol and purpose of the study, then the child's parent or guardian must consent. If the child or adolescent is old enough to understand the purpose of the research and what is being asked, then the child/adolescent gives *assent* and the parents or guardian is still required to give *consent*. However, the child's wishes override those of his or her parent: If the child/adolescent does not want to participate, his or her wishes hold, no matter what the parent desires.

Does this requirement for consent mean that research cannot be conducted in public places? The general rule is that researchers must be careful to inform persons they are studying whenever possible, and if they are conducting studies in a public place, they should not collect identifying data without permission. For instance, if you are conducting a study on a beach, you might announce the study in the local newspaper, place signs on lampposts in the area, and request agreement from anyone approached directly. But because you may observe a number of individuals who have not given their consent to participate, you should not record any information that identifies particular participants.

If you think that participants may object to having their likenesses (in photographs or videos) appear in the public domain, but will permit their use for research purposes, you can split the consent form into two parts to get participants for your study, so that the prospect of the public use of visual media won't interfere with recruitment. Participants may sign one or both sections (see Figures 11.2 and 11.3 for examples).

⧚ SUMMARY

Before actually beginning your study, you must do considerable work. You must prepare the proposal, outlining what you intend to do, where, and why; and you must obtain approvals, including from those responsible for providing ethics approval and from those with authority over your research site.

We have noted in this chapter some of the requirements for ethics review committee approval of your proposal, such as information on what will be required of participants and how you intend to use the information obtained. We have also listed the rights of participants, which include the right to be protected from harm and the right to be informed about the research aims.

[Title of your project]

[Your name, affiliation, contact phone number]

[A brief description of your project: Say what will be required of participants, and approximately how long it will take. If you are providing participants with reimbursement, that should be stated here.]

Consent

I hereby consent to participate in the above research project. I understand that my participation is voluntary and that I may change my mind and refuse to participate or withdraw at any time without penalty. I may refuse to answer any questions or I may stop the interview. I understand that some of the things that I say may be directly quoted in the text of the final report and subsequent publications, but my name will not be associated with that text.

I hereby agree to participate in the above research:

_____	_____	_____
Participant	Print Name	Date
_____	_____	_____
Principal Investigator	Print Name	Date
_____	_____	_____
Witness	Print Name	Date

Figure 11.2   Sample of a Consent Form

I hereby give consent to be photographed for this research. I understand that my name will not be associated with these photographs. The photographs may be published with the final report and/or articles in professional journals or used for educational purposes.

_____	_____	_____
Participant	Print Name	Date
_____	_____	_____
Principal Investigator	Print Name	Date
_____	_____	_____
Witness	Print Name	Date

Figure 11.3   Sample of an Addendum to a Consent Form for a Study Using Photographs or Videotapes

⧫ RE/OURCE/

On Writing Proposals

Boyd, C. O., & Munhall, P. L. (2001). Qualitative proposals and reports. In P. L. Munhall (Ed.), *Nursing research: A qualitative perspective* (3rd ed., pp. 613–638). Boston: Jones & Bartlett.

Morse, J. M. (2003). A review committee's guide for evaluating qualitative proposals. *Qualitative Health Research, 13,* 833–851.

Morse, J. M., & Field, P. A. (1995). *Qualitative research methods for health professionals* (2nd ed.). Thousand Oaks, CA: Sage.

On Preparing Proposals for Funding

Cheek, J. (2000). An untold story? Doing funded qualitative research. In N. K. Denzin & Y. S. Lincoln (Eds.), *Handbook of qualitative research* (2nd ed., pp. 401–420). Thousand Oaks, CA: Sage.

Morse, J. M. (1994a). Designing funded qualitative research. In N. K. Denzin & Y. S. Lincoln (Eds.), *Handbook of qualitative research* (pp. 220–235). Thousand Oaks, CA: Sage.

On Ethics Requirements and Approvals

Christians, C. G. (2000). Ethics and politics in qualitative research. In N. K. Denzin & Y. S. Lincoln (Eds.), *Handbook of qualitative research* (2nd ed., pp. 133–155). Thousand Oaks, CA: Sage.

Mauthner, M., Birch, M., Jessop, J., & Mueller, T. (Eds.). (2002). *Ethics in qualitative research.* Thousand Oaks, CA: Sage.

General Resources

Finch, J. (1986). *Research and policy: The uses of qualitative research in social and educational research.* London: Falmer.

Kvale, S. (1996). *InterViews: An introduction to qualitative research interviewing.* Thousand Oaks, CA: Sage.

Marshall, C., & Rossman, G. B. (1999). *Designing qualitative research* (3rd ed.). Thousand Oaks, CA: Sage.

Miller, D. C. (1991). *Handbook of research design and social measurement* (5th ed.). Newbury Park, CA: Sage.

Moustakas, C. (1990). *Heuristic research: Design, methodology, and applications.* Newbury Park, CA: Sage.

Piantanida, M., & Garman, N. B. (1999). *The qualitative dissertation: A guide for students and faculty.* Thousand Oaks, CA: Sage.

Van Maanen, J., Dabbs, J. M., & Faulkner, R. R. (1982). *Varieties of qualitative research.* Beverly Hills, CA: Sage.

12

Getting Started

Qualitative research is only as good as the investigator. This is both the good and the bad news. It is the researcher who makes all the analytic decisions, not the data, not the method, not the computer. It is the researcher who, through skill, persistence, patience, and wisdom, earns the trust of participants in the setting, flexibly adapts the strategy, and elegantly balances the design. It is the researcher who makes the necessary data to produce a rich study, ensures methodological congruence, and is meticulous about documentation. It is the researcher who is well prepared about the topic, incisively interrogates data, accurately recognizes clues, and sensitively interprets. This is a craft with high standards.

Learning any craft is made easier by knowing the standards of how it is done when it is done well. Good researchers set high standards. They are familiar with the relevant theory and studies, so they recognize what is known and what is not known, and they separate these things from those that are new, puzzling, and significant. They can locate an appropriate framework or paradigm to work within, and they have the ability to keep this apart from the data so that inductive work is still possible. They are stimulated, not stopped, by ambiguity or paradoxes, but have the ability to recognize that reality is not simple; rather, it is a complex puzzle, a challenge, to which there may be no best and easy answer. The intellectual work of making sense of their data is a challenge they revel in, because they know the research methods, strategies, and techniques they need. They also know these are only tools to facilitate inquiry— means to the end, not the solution itself. Researchers use strategies and techniques to manipulate data so that they may be analyzed; strategies and techniques themselves do not do the analysis.

How, then, do you become a good researcher? The description above might seem daunting for a novice, and it might even be the reason you hesitate to start. But like any craft, qualitative research is learned by doing it. You become a qualitative researcher in the process of doing research, learning from your mistakes as well as from your successes. The concern is not how to become a researcher but how to learn best while doing research.

Qualitative research will not mysteriously happen by itself. No project ever self-started. Throughout the research process, the researcher is constantly making choices, asking questions of the data, and asking questions outside of the data. That process starts when you make it start. The corollary, of course, is that it is very easy to postpone starting. So we conclude this book with a focus on this entirely achievable task.

◊ WHY Iʃ IT ʃO HARD TO ʃTART?

If getting started is a problem, it helps to see that this is built into the method. Getting started in any research project, especially a significant one, is always difficult. But qualitative research offers particular challenges that are much easier to deal with if you know they are not your fault.

First, although in a sense qualitative research, like quantitative research, begins when you start thinking about the topic, it *actually begins* when you first enter the setting and conduct your first observation or first interview. The hardest part is walking up to the first participant and explaining your study in two sentences or less. Prepare for this by role-playing and by practicing with your colleagues.

Second, in qualitative research, everything seems to happen at once. At least two things *must* happen simultaneously: the making of data and the exploration of data. This is because the methods are data driven. What you come to understand from early field research or interviewing may change your framing of the research question and hence demand changes in the data-construction process. Thus, as data records accrue, you must explore them, code them, discuss them with team members or participants, and question them in the light of the literature, and then, in turn, you must explore and record the results of these explorations. So wherever you begin, several processes start at once.

Third, qualitative modes of data collection can be confusing and demanding. Getting accepted as an observer in a strange setting or making it through a first unstructured interview can be nerve-racking even for the most experienced observer or interviewer. For a novice, these first steps can seem a sufficient challenge without the additional need to record impressions, code new discoveries, and diagram connections. These techniques of making data also can have a formidable momentum, making withdrawal and reflection very difficult.

ⅲ HOW DO YOU START?

So how do you start? This book is full of answers to that question. If you are ready to start, review them. Determined to discover understanding from the research site or participants, a researcher must begin somewhere.

Start in the Library

Judging whether your research will produce new knowledge involves knowing what's known. There is no alternative to reading extensively and widely on the selected research topic. Start with what's known.

We have argued throughout this book against the myth that prior literature searching and prior theorizing are inimical to theory discovery. If something is known, there is absolutely no point in denying yourself that knowledge, and every reason to suppose your study will be better focused and more useful if informed by prior knowledge. Start by finding out what's known and what is still being asked about your research area. Focus on *local* knowledge: What's known about the setting you are studying, the area you are in? Start building up more general knowledge of previous research on the broader topic—but *don't* wait until your knowledge is complete to start.

Starting in the library does not mean you are delaying the research. As you read, treat the outcome of your reading—notes, reviews, memos—as data. These are data records, to be coded and explored along with your interview transcripts or memos. If you are working on the computer, use your qualitative software to analyze literature searches and notes as data as well as alongside your data.

A literature *review* requires techniques similar to those of qualitative analysis—distilling a lot of messy data and finding their story, patterns, or themes. Like most qualitative research, it will start with some questions and generate more. Your first question is "What is known?" As your review progresses, you will discover more specific puzzles ("Why do American studies have results so different from European ones?" or "What theory would be needed to account for this aspect of the topic, and why has it not been developed?").

Reading extensively on the topic is compatible with working inductively. As you read, work on the goal of bracketing knowledge. Good data management will assist with the bracketing process.

⟩⟩⟩ Always identify material from the literature (type it in a different font, or on colored paper, or code it at a "literature" category). Keep the literature, your ideas about the literature, and any preconceived notions you have about the topic in separate storage areas, each separate from the data.

⟩⟩⟩ Attach authors' names to ideas, concepts, and theories as a constant reminder of whose ideas are being discussed. Code each piece of text to its citation.

⟩⟩⟩ As you read, record reminders of preliminary thinking to inform your account of the process of analysis. Remain a skeptic toward knowledge you find in your reading, and test every assumption.

Start With an Armchair Walkthrough

Reread the ways of systematically thinking out your topic through methodological options. What would your study look like and feel like if it were done by each of the methods sketched there? How would you focus it on a research question, and what sort of data would you seek? Even if, or especially if, you approach this topic committed to a particular method, such a mindful stroll may open possibilities and help firm a research design.

⟩⟩⟩ Record your walkthrough by taping your thinking aloud, drawing it, or noting it in matrix form, as we have shown in Chapter 2.

⟩⟩⟩ Start memos immediately. Projects begin with ideas, not data. Develop a memo-writing routine that you like and will use easily, and record in memos your first thoughts about the project, the appropriate method, the possible routes to take, and the reasons you are studying this topic.

〰 If you will be working on the computer, start using your software. Store the record of your walkthrough as your first project document and the memos recording your thoughts as more documents.

Start Thinking Method

The walkthrough is a first step toward thinking your way into the appropriate method. It is not a substitute for thorough reading of texts and examples of project reports. Start with the list of references at the end of this book, and branch out from there to immerse yourself in the method and the sort of research it supports. Seek out and *critique* dissertations, books, and conference papers that report on studies using that method.

〰 Avoid becoming a methodological cultist: Read outside the method and listen to criticisms of it. From the outset, be alert to its challenges as well as its advantages for your project.

〰 If possible, find a researcher skilled in the method you are studying and work with him or her, learning as an apprentice does, by observing and following the lead until you can pick up the way to think about data and analysis.

Start With Yourself

Should you include data from personal experience in the study? As we noted in Chapter 5, it is sometimes argued (and was argued strenuously in early anthropology) that researchers should separate themselves from the topics and the people they study to avoid driving their research problems with their own personal agendas. It has been feared that personal involvement makes the topic more stressful, that it leads to researchers' losing their "objectivity" and to less fairness in reporting. Anthropologists argued that researchers cannot "see" a particular culture's values and beliefs if they are immersed in that culture. These objections have now by and large been relaxed, but, as Lipson has noted (see "Dialogue," 1991), this task is not easy and takes concerted effort.

Where do you place your own experience when your values, beliefs, culture, and even physical limitations affect the process and quality of data? As you read the literature where your project is located, you will find that most researchers are concerned with this question. Adherents of one school of thought aim for objectivity by depersonalizing the data, emphasizing issues of reliability and validity in data collection and considering those being studied as "informants" or "actors." The extreme alternative is to acknowledge the researcher as a part of the setting so that his or her interpretation is just one among many, including those of participants—or even coresearchers—in the research process. The researcher must consciously choose, negotiate, and maintain the relationship he or she has with the persons being studied.

What Role Should the Researcher's Personal Experience Play?

In Chapter 5, we discussed the pros and cons of the researcher's using his or her personal experience or incorporating that experience into the project. We noted that the researcher may use his or her own experience in two ways: by delineating it and using it as data while separating it from experiences of others in the study, or by using it as legitimate and intimately rich data that are perhaps more valid than the reported, secondhand experiences obtained by interviewing or observing others. Whatever your choice, we advise you to use your experience with care.

Hidden Agendas

Along with the advantages that come with using your own experience, such as making the study richer, there are dangers, and words of caution are needed. Researchers often have hidden agendas—that is, they may have issues they are seeking to resolve and may select these issues as topics for study. Consider the possibility that selecting a problem in which you have personal involvement may make the research project an overwhelmingly distressing experience. Recall that qualitative data collection requires listening to similar stories of those who have also been through the experience, and then dwelling with those stories for months. Do you want to do this to yourself, especially if the material is personally and profoundly disturbing? We do not recommend that you try to understand your own experience of a disturbing event, such as the death of a parent, by conducting a qualitative study.

If you have an ax to grind, similar advice holds. Resolve your problems using the ordinary channels. If you feel you were given a speeding ticket unfairly, do not select attitudes of traffic police toward drivers as your topic. If you feel you were treated unfairly in some other arena, do not attempt to resolve your problem with qualitative research. Above all, you must remember that there is a clear distinction between detective work and qualitative research. You belong with the latter—qualitative research is not a "gotcha" endeavor. Stay clear of high-risk topics that may result in encounters with the law, the subpoenaing of data, and personal risk.

Occasionally in contract research, the contracting agency will have a hidden agenda and will use the qualitative researcher to help make a point, justify its existence, and so forth. The researcher may encounter problems when the agency's agenda is never made explicit, especially when the researcher's results do not meet the agency's expectations. The researcher can avoid such problems, at least in part, if the contracting agreement is clear on certain points, such as to whom the data belong, the limits of confidentiality of the findings and of the host organization, and the publication rights of the researcher.

In summary, you should choose your research topic with care, because selecting something for the wrong reason may result in unnecessary distress and/or poor research. Making your selection may involve some self-reflection ("Why does this topic grab me?") and an assessment of gains or costs to self. Know yourself.

Start Small

Researchers are easily swamped by qualitative data. As in surfing, the challenge is to get on top and stay on top as the volume and momentum of the wave of data build up. When data tower above you, it's too late to get on top. The momentum of qualitative research can be formidable, because modes of sampling and data acquisition are often not under the researcher's control. Once accepted in a fieldwork setting, you are swept up in the events you must observe. Once started on a snowball sample, you can hardly refuse offers of further interviews.

There are serious problems for a project using any qualitative method if the researcher starts data collection without starting analysis, thus allowing data to build up untouched. In 1989, Anselm Strauss memorably described this situation to Lyn Richards: "Students get data crawling up their backs." Qualitative data are, by their nature, rich, complex, and defiant of instant reduction to tidy shapes or numbers. Rapidly acquiring many records without exploring and coding them can be disastrous to the researcher's sense of direction and access to the data. We have offered advice on ways of staying on top and processing data records as they come into the project. As for getting started, our advice is simple: Start small.

ℳ Start with a limited, relatively self-contained project segment, perhaps with a marginally relevant group, or a documentary analysis of setting the scene. If things go wrong, you can withdraw tactfully without damaging the main project.

ℳ From the start, review, consider, and in some way process each item of data as it is brought into the project. On reading it, write a memo recording your memories of the event, first impressions of the interview, and the like.

Start Safe

Maintain control over the research process by strategically planning your entry into the research field or your first interviews. Most projects offer some choice as to where the research starts, and you can use this choice to delay research events that are particularly crucial or encounters about which you are not confident.

ℳ Practice interviewing with friends or family before doing interviews in your research field.

ℳ Think out carefully the ethical and courtesy aspects of each research initiative, and consult in the field as widely as possible.

ℳ If possible, start in consultation with research confidants or advisers who will talk out with you what happened and what you should do about it.

Start Soon

Delaying the start of a project can have very serious effects. Apart from the impact of delaying on your self-image and deadline, it can lead to a loss of momentum and failure to attain the knowledge level you had planned to reach. The methodological foreplay of reviewing literature, planning, replanning, and consulting is absolutely necessary to any project, but it should not be allowed to take over the research. There is always somewhere you can sensibly start.

Start with caution, especially if you are under pressure to produce results. If clients or supervisors simultaneously demand too much, give them a copy of this book, and while they read it, get on with placing the project, becoming skilled, and designing your entry into the field. These processes need not take long, but you must do them. Before the client has finished this book, you will be ready. Then . . .

Start With a Research Design

Locating a research question and selecting a method do not provide you with an instant research design. You can't start without a practical outline of a doable project. This is so important that we have devoted an entire chapter to it (see Chapter 4). But you *can* start a project without first making your research design perfect. And to start it, you must commence with careful consideration of the knowledge you are bringing in, which is the first step in building a good relationship with your research project, a sense that it can be done, that there is a task here you can get your arms around.

Start Skilled

From your armchair walkthrough and your detailed reading on the method to be used, note any skills you will need that you do not yet have and note *when* you will need them. Make it a rule not to commence any stage of the project before you are comfortable with the required skills. If you do not have the research experience with the methods of data collection that your project requires, be sure to learn more about those methods before you begin data collection. Start with the resources listed in this book and read widely on particular methods of making or analyzing data, so that you know not only what they entail but also how to know if you are doing them well.

Never test data-collection techniques in real research situations. All qualitative data making is potentially invasive and intrusive, so your mistakes may cause someone else distress. Qualitative data are also cumulative, and informants are likely to communicate, so any mistakes you make are also likely to impede later data-collection efforts.

Start in Your Software

Choose your software early and become competent in computer use and all the relevant software *before* data collection begins. If you wait until you are making data, you will risk damaging or even losing files.

◊ If you are not confident with computers, seek help to ensure that you can manage the operating system and do good housekeeping of your files and backups. You need those skills for any research work, in any software.

◊ Ensure that you are competent in all the software you will be using, including your word-processing software (Can you format and edit well?).

◊ Now, your qualitative software. You need basic competence in it from the start of research design so that the project can benefit from software from the start. Assess your ability to self-teach. Good software will include self-teaching materials, and you can use the tutorials outlined in Appendix 1.

◊ Use others' wisdom! Don't assume self-teaching is best: Seek colleagues or helpers who have the software skills and experience you want. Find what software support is available in your institution and use it. Go to the software maker's website to find workshops, consultants, or virtual courses. Seek out Internet discussion lists devoted to the software, so you can pick up tips from other researchers and avoid their mistakes. Learning with others is usually far more productive and often faster and much more fun than learning alone.

◊ CONGRATULATIONS, YOU'VE STARTED!

Do you remember why you wanted to do qualitative research? Keep this and your project goals in mind as you work, and remain malleable and open to your participants and to your data. Do justice to the richness of your data. And above all, have fun.

◊ RESOURCES

Cole, A. L., & Hunt, D. E. (1994). *The doctoral thesis journey: Reflections from travellers and guides.* Toronto: Ontario Institute for Studies in Education.
Fitzpatrick, J., Secrist, J., & Wright, D. J. (1998). *Secrets for a successful dissertation.* Thousand Oaks, CA: Sage.

King, N. M. P., Henderson, G. E., & Stein, J. (1999). *Beyond regulation: Ethics in human subjects research.* Chapel Hill: University of North Carolina Press.

Locke, L. F., Spirduso, W. W., & Silverman, S. J. (1993). *Proposals that work: A guide for planning dissertations and grant proposals* (3rd ed.). Newbury Park, CA: Sage.

Meloy, J. M. (1994). *Writing the qualitative dissertation: Understanding by doing.* Hillsdale, NJ: Lawrence Erlbaum.

Richards, L. (2005). *Handling qualitative data: A practical guide. Part I: Setting up.* London: Sage.

Rossman, M. H. (1995). *Negotiating graduate school: A guide for graduate students.* Thousand Oaks, CA: Sage.

Rudestam, K. E., & Newton, R. B. R. (1992). *Surviving your dissertation: A comprehensive guide to content and process.* Newbury Park, CA: Sage.

Sieber, J. E. (1992). *Planning ethically responsible research: A guide for students and internal review boards.* Newbury Park, CA: Sage.

Appendix 1

Using the Software Tutorials

Lyn Richards

We have designed this book to give beginning researchers an understanding of what it is like to work qualitatively. An important part of this experience now is working qualitatively on the computer. Although we do not assume that all qualitative researchers today will use computers, many texts and most employers now do make that assumption, and the researcher who knows what specialized software can offer is greatly helped in assessing what he or she can achieve. Researchers approaching qualitative software for the first time often find it daunting, but in our experience, their problem is usually simply that they do not know what it would be like to work with qualitative software.

To help you find out, this Appendix takes you to simple tutorials that offer a quick picture of the way one package, QSR NVivo, supports qualitative research. You can go through the tutorials using your own data, if you already have data, or use the files provided with the free demonstration version of the software.

The previous edition of this book carried a CD-ROM with demonstration software and tutorials. This second edition links instead to the tutorials on the Web, since (unlike a CD in a book) they can be regularly updated.

Author's Note: Tutorials are on the website at: http://www.sagepub.co.uk/richards

The tutorials and instructions for accessing the software are provided on the Web at http://www.sagepub.co.uk/richards/. They are designed to assist readers of either this book or my more detailed text on handling data (Richards, 2005). You can use the tutorials from the website or download them in .pdf format.

Return to Chapter 4 to review what is offered by software for qualitative research, and to find up-to-date reviews of current software. NVivo and its partner program, N6, are two of these programs, developed by Tom Richards and myself in a university research project that acquired, by accident, the acronym NUD*IST (from an observation that researchers with this sort of Nonnumerical, Unstructured Data require ways of Indexing, Searching, and Theorizing). NVivo is developed and distributed by QSR International: http://www.qsrinternational.com.

⧵⧵ USING THE TUTORIALS WITH THE CHAPTERS IN THIS BOOK

Chapters 4 to 7 of this book introduced the approaches available with software, the advances they offer, and challenges to which you should be alerted. Each of those four chapters can be read alongside two of the online software tutorials.

Research Design (Chapter 4)

In Chapter 4, we stress the importance of managing the information and ideas for your project from its inception, and how to use software during this critical first stage, to log your thinking around the research design and help clarify the choices you have and the decisions you make. We urge starting immediately with recording ideas and proposals early, so that you can track the genesis of the project.

Working in software, this can be done by creating the project that will hold all your data and ideas, structuring it to manage them well, and beginning with early planning documents.

To follow these steps in the online tutorials, work through Tutorial 1: *Setting Up Your Project*. This contains an introduction to the software and how to use its online Help files, and then instructions to create a project.

Making Data (Chapter 5)

Chapter 5 emphasizes the many ways of creating data records. To do this in software, you need to be able to make documents in your project or import them from word processor files. Among the advances software supports are the ability to store information about the people or places you are studying, to annotate and link your documents, to write memos about them, and to make links to nontext data.

To follow these steps in the online tutorials, work through the next three tutorials:

❧ Tutorial 2: *Creating and Importing Sources*

❧ Tutorial 3: *Managing Data: Cases, Attributes, and Sets*

❧ Tutorial 4: *Editing and Linking: Getting Up From the Data*

Coding (Chapter 6)

In Chapter 6, we describe the many ways of coding and their different purposes in qualitative research. One of these is descriptive coding by storing values of attributes, and you learned how to do this in software in Tutorial 3. The two forms of qualitative coding—topic coding and analytical coding—require different software skills.

To learn how to do these in NVivo, follow the steps in Tutorial 5: *Coding, and Working With Coded Data.*

Abstracting (Chapter 7)

As the project grows, you will make many more categories, and they will become more abstract and more interesting. As we argue in Chapter 7, managing these categories well greatly enhances the analysis stage.

❧ How to do this in NVivo is explained in Tutorial 6: *Relationships and Other Nodes: Handling Ideas.*

And as the ideas take shape, you may wish to use models or displays to help you abstract. For instructions in doing this, go to Tutorial 7: *Seeing It in Models.*

Getting It Right, and Writing (Chapters 9 and 10)

What next? Software offers ways of querying your data and testing your hunches that were not available to manual methods. As you work toward the goals later in this book, "getting it right," these tools will come into play. To explore them, complete the final tutorials on the website:

⦚ Tutorial 8: *Finding Items and Querying the Data*

⦚ Tutorial 9: *Exploring Patterns in Matrices*

⦚ Tutorial 10: *Reporting and Showing Your Project*

These tutorials are designed to give you a feel for how it is to work with software rather than to instruct you on doing a project in NVivo. A strong argument we make throughout this book is that qualitative research involves active handling of data. In learning software tools like coding and searching, always be aware of the researcher's agency. The computer can code automatically, but the researcher must point it at what is to be coded where. The computer can do very sophisticated searches, scoping them exactly and delivering the results, but the researcher must design the search and think about the results. Doing coding is easy, but using the interpretation it expresses is a skilled task. Doing searching is rapid and dramatic, but using searches requires you to think like a researcher.

Should you use this software in your own research, with practice, you will reach for the relevant tool when you want to store information or interpretation, or to ask a question. (But never forget that it is you, the researcher, who chooses what is to be stored, does the interpreting, and asks the questions!)

If you wish to continue exploring the software, with your own data or the data provided, move now from the tutorial mode to focus on the data. Using just the tools learned in these tutorials, you could conduct a project to completion. As you read and reflect, reach for the relevant tool to do what is needed according to the method you are working with, the design you have created, the discoveries you are making, and the questions you are asking. Approached this way, software becomes a resource for the researcher, never a substitute for the research method and the research mind.

Appendix 2

Applying for Funding

Janice M. Morse

Excellent qualitative inquiry takes time. Excellent qualitative inquiry must be built on the foundations of literature and the work of others and fit into what is known; therefore a research assistant can facilitate many labor-intensive tasks. Excellent qualitative inquiry requires good data, careful transcription, consultation for experienced researchers, and assistance from more junior researchers, which is often expensive. Excellent qualitative inquiry may require equipment (computers and printers, tape recorders and transcribers, software, cameras, and copy machines), and this equipment is not always available and must be purchased. Good qualitative inquiry must be disseminated at conferences; participants need to be informed of the results, which then must be prepared and submitted for publication. In short, good qualitative inquiry is expensive to do and expensive to disseminate.

Obviously, if the research can obtain a grant to assist with or cover these expenses, then the researcher's task of doing research is lightened and the chances for success improved. But obtaining a grant is an art, and knowing the tricks of grantsmanship will greatly enhance your chances of success! This appendix offers some advice.

≫ APPLYING FOR FUNDING

Every granting agency has "terms of reference" or a statement of goal and priority topics describing the type of areas it is interested in funding. It is wise to double-check that your project would be eligible for funding by looking at the lists of projects previously funded to determine that yours is similar. If still in doubt, then phone the agency and ask. While doing so, check the following:

1. Do they fund researchers at your level of expertise? (Do they fund students? New researchers without a track record?)

2. How much funding will they provide? A researcher may require only a small "seed grant," enough to pay for assistance with transcription and other minor costs, or a larger grant that will also support equipment and staff over a number of years.

3. Do they consider qualitative research? (Check first the titles of previously funded grants.)

Pay careful attention to the application deadlines and other agency requirements, allowing adequate time for the necessary signatures and approvals from your own institution. Note that the grant is normally provided to your institution, not to you as an individual. If you are a graduate student, then your advisor will usually be required to sign the grant as principal investigator.

Can you apply for funding from more than one agency simultaneously? Usually this is acceptable, as long as you notify each agency that you have also submitted elsewhere and name the other agency(ies). Then, if your proposal is rated in the fundable range, the agency will contact the other funding body, and perhaps both agencies will share the costs.

≫ ONCE FUNDED

Once the grant is funded, your timeline begins. Be certain to fulfill all of the agency's requirements, including submitting interim and final reports as requested.

Qualitative research, even excellent qualitative work, has an uncertain course and outcome at best. Communicate regularly with the agency's internal research officer responsible for your grant—especially if you deviate from your proposal. Get official permission if you want to make changes to your proposed design. And, once you complete your work, be certain to acknowledge your funding sources in all presentations and publications!

References

Agar, M. H. (1986). *Speaking of ethnography.* Beverly Hills, CA: Sage.

Agar, M. H. (1996). *The professional stranger: An informal introduction to ethnography* (2nd ed.). San Diego, CA: Academic Press.

Alasuutari, P. (1995). *Researching culture: Qualitative method and cultural studies.* London: Sage.

Albrecht, G. L. (1985). Videotape safaris: Entering the field with a camera. *Qualitative Sociology, 8*(4), 325–344.

Altheide, D. L., & Johnson, J. M. (1994). Criteria for assessing interpretive validity in qualitative research. In N. K. Denzin & Y. S. Lincoln (Eds.), *Handbook of qualitative research* (pp. 485–499). Thousand Oaks, CA: Sage.

Angrosino, M. V., & Mays de Pérez, K. A. (2000). Rethinking observation: From method to context. In N. K. Denzin & Y. S. Lincoln (Eds.), *Handbook of qualitative research* (2nd ed., pp. 673–702). Thousand Oaks, CA: Sage.

Applegate, M., & Morse, J. M. (1994). Personal privacy and interaction patterns in a nursing home. *Journal of Aging Studies, 8,* 413–434.

Atkinson, P., & Hammersley, M. (1994). Ethnography and participant observation. In N. K. Denzin & Y. S. Lincoln (Eds.), *Handbook of qualitative research* (pp. 248–261). Thousand Oaks, CA: Sage.

Atkinson, R. (1997). *The life story interview.* Thousand Oaks, CA: Sage.

Ball, M. S., & Smith, G. W. H. (1992). *Analyzing visual data.* Newbury Park, CA: Sage.

Bates, G., & Mead, M. (1942). *Balinese character: A photographic analysis.* New York: New York Academy of Sciences.

Bazeley, P. (1999). The bricoleur with a computer, piecing together qualitative and quantitative data. *Qualitative Health Research, 9,* 279–287.

Bazeley, P. (2003). Computerized data analysis for mixed methods research. In A. Tashakkori & C. Teddlie (Eds.), *Handbook of mixed methods in social and behavioral research* (pp. 385–422). Thousand Oaks, CA: Sage.

Bazeley, P. (in press). The contribution of computer software to integrating qualitative and quantitative data and analyses. *Research in the Schools.*

Bazeley, P. (forthcoming, 2007). *Qualitative Analysis with NVivo.* London: Sage.

Bazeley, P., & Richards, L. (2000). *The NVivo qualitative project book.* London: Sage.

Becker, H. S. (1986). *Writing for social scientists: How to start and finish your thesis, book, or article.* Chicago: University of Chicago Press.

Benner, P. (Ed.). (1994a). *Interpretive phenomenology: Embodiment, caring, and ethics in health and illness.* Thousand Oaks, CA: Sage.

Benner, P. (1994b). Preface. In P. Benner (Ed.), *Interpretive phenomenology: Embodiment, caring, and ethics in health and illness.* Thousand Oaks, CA: Sage.

Berg, B. L. (1998). *Qualitative research methods for the social sciences* (3rd ed.). Boston: Allyn & Bacon.

Bernard, H. R. (1988). *Research methods in cultural anthropology.* Newbury Park, CA: Sage.

Bernard, H. R. (1994). *Research methods in anthropology: Qualitative and quantitative approaches.* Walnut Creek, CA: AltaMira.

Bernard, H. R. (2000). *Social research methods: Qualitative and quantitative approaches.* Thousand Oaks, CA: Sage.

Beverley, J. (2000). Testimonio, subalternity, and narrative authority. In N. K. Denzin & Y. S. Lincoln (Eds.), *Handbook of qualitative research* (2nd ed., pp. 555–565). Thousand Oaks, CA: Sage.

Blumer, H. (1986). *Symbolic interactionism: Perspective and method.* Berkeley: University of California Press. (Original work published 1969)

Bottorff, J. L. (1994). Using videotaped recordings in qualitative research. In J. M. Morse (Ed.), *Critical issues in qualitative research methods* (pp. 244–261). Thousand Oaks, CA: Sage.

Boyatzis, R. E. (1998). *Transforming qualitative information: Thematic analysis and code development.* Thousand Oaks, CA: Sage.

Boyd, C. O. (1993). Phenomenology: The method. In P. L. Munhall & C. O. Boyd (Eds.), *Nursing research: A qualitative perspective* (2nd ed., pp. 99–132). New York: National League for Nursing.

Boyd, C. O., & Munhall, P. L. (2001). Qualitative proposals and reports. In P. L. Munhall (Ed.), *Nursing research: A qualitative perspective* (3rd ed., pp. 613–638). Boston: Jones & Bartlett.

Boyle, J. S. (1994). Styles of ethnography. In J. M. Morse (Ed.), *Critical issues in qualitative research methods* (pp. 159–185). Thousand Oaks, CA: Sage.

Boyle, J. S. (1997). Writing it up: Dissecting the dissertation. In J. M. Morse (Ed.), *Completing a qualitative project: Details and dialogue* (pp. 9–37). Thousand Oaks, CA: Sage.

Brizuela, D., Stewart, J. P., Carrillo, R. G., & Garbey, J. (2000). *Acts of inquiry and qualitative research.* Cambridge, MA: Harvard Educational Review.

Bryman, A. (2006). Integrating quantitative and qualitative research: How is it done? *Qualitative Research, 6*(1), 97–113.

Carey, M. A. (Ed.). (1995). Issues and applications of focus groups. *Qualitative Health Research, 5,* 413–524.

Carney, T. F. (1990). *Collaborative inquiry methodology.* Windsor, ON: University of Windsor, Division of Instructional Development.

Carspecken, P. F. (1996). *Critical ethnography in educational research.* New York: Routledge.

Cassell, J. (1992). On control, certitude and the "paranoia" of surgeons. In J. M. Morse (Ed.), *Qualitative health research* (pp. 170–191). Newbury Park, CA: Sage. (Original work published 1987)

Charmaz, K. (2000). Grounded theory: Objectivist and constructivist methods. In N. K. Denzin & Y. S. Lincoln (Eds.), *Handbook of qualitative research* (2nd ed., pp. 509–535). Thousand Oaks, CA: Sage.

Charmaz, K. (2006). *Constructing grounded theory: A practical guide through qualitative analysis.* Thousand Oaks, CA: Sage.

Cheek, J. (2000). An untold story? Doing funded qualitative research. In N. K. Denzin & Y. S. Lincoln (Eds.), *Handbook of qualitative research* (2nd ed., pp. 401–420). Thousand Oaks, CA: Sage.

Chenitz, W. C., & Swanson, J. M. (1986). *From practice to grounded theory.* Reading, MA: Addison-Wesley.

Christians, C. G. (2000). Ethics and politics in qualitative research. In N. K. Denzin & Y. S. Lincoln (Eds.), *Handbook of qualitative research* (2nd ed., pp. 133–155). Thousand Oaks, CA: Sage.

CIHR/NSERC/CRSNG/SSHRC. (2001). Section 1: Ethics review. In *Tri-council policy statement: Ethical conduct for research involving humans.* Retrieved September 8, 2001, from http:www.nserc.ca/programs/ethics/english/sec01.htm

Clarke, A. (2005). *Situational analysis: Grounded theory after the postmodern turn.* Thousand Oaks, CA: Sage.

Clarke, M. (1992). Memories of breathing: Asthma as a way of becoming. In J. M. Morse (Ed.), *Qualitative health research* (pp. 123–140). Newbury Park, CA: Sage. (Original work published 1990)

Coffey, A., & Atkinson, P. (1996). *Making sense of qualitative data.* Thousand Oaks, CA: Sage.

Coffey, A., Holbrook, B., & Atkinson, P. (1996). Qualitative data analysis: Technologies and representations. *Sociological Research Online, 1*(1). Retrieved May 8, 2001, from http://www.socresonline.org.uk/socresonline/1/1/4.html

Cole, A. L., & Hunt, D. E. (1994). *The doctoral thesis journey: Reflections from travellers and guides.* Toronto: Ontario Institute for Studies in Education.

Corbin, J., & Strauss, A. L. (1992). A nursing model for chronic illness management based on the trajectory framework. In P. Woog (Ed.), *The chronic illness trajectory framework: The Corbin and Strauss nursing model* (pp. 9–28). New York: Springer.

Couch, C. J. (1986). Questionnaires, naturalistic observations, and recordings. In C. J. Couch, M. A. Katovich, & S. L. Saxton (Eds.), *Studies in symbolic interaction, supplement 2: The Iowa school (Part A)* (pp. 45–59). Greenwich, CT: JAI.

Crabtree, B. F., & Miller, W. L. (Eds.). (1999). *Doing qualitative research* (2nd ed.). Thousand Oaks, CA: Sage.

Creswell, J. W. (1994). *Research design: Qualitative and quantitative approaches.* Thousand Oaks, CA: Sage.

Creswell, J. W. (1998). *Qualitative inquiry and research design: Choosing among five traditions.* Thousand Oaks, CA: Sage.

Creswell, J. W. (2003). *Research design: Qualitative, quantitative and mixed method approaches* (2nd ed.). Thousand Oaks, CA: Sage.

Davis, D. L. (1983). *Blood and nerves: An ethnographic focus on menopause.* St John's: Memorial University of Newfoundland.

Davis, D. L. (1992). The meaning of menopause in a Newfoundland fishing village. In J. M. Morse (Ed.), *Qualitative health research* (pp. 145–169). Newbury Park, CA: Sage. (Original work published 1986)

Davis, D. S. (1991). Rich cases: Ethics of thick description. *Hastings Center Report, 21*(4), 12–17.

Daymon, C., Holloway, I., & Daymon, C. (2002). *Qualitative research methods and public relations & marketing communications.* London: Routledge.

Denzin, N. K. (1989). *Interpretive biography.* Newbury Park, CA: Sage.

Denzin, N. K. (1997). *Interpretive ethnography: Ethnographic practices for the 21st century.* Thousand Oaks, CA: Sage.

Denzin, N. K., & Lincoln, Y. S. (Eds.). (1994). *Handbook of qualitative research.* Thousand Oaks, CA: Sage.

Denzin, N. K., & Lincoln, Y. S. (Eds.). (1998). *The landscape of qualitative research: Theories and issues.* Thousand Oaks, CA: Sage.

Denzin, N. K., & Lincoln, Y. S. (Eds.). (2000). *Handbook of qualitative research* (2nd ed.). Thousand Oaks, CA: Sage.

Denzin, N. K., & Lincoln, Y. S. (Eds.). (2005). *The Sage handbook of qualitative research* (3rd ed.). Thousand Oaks, CA: Sage.

Dewalt, K. N., & Dewalt, B. R. (2002). *Participant observation: A guide for field workers.* Walnut Creek, CA: AltaMira.

Dey, I. (1995). *Qualitative data analysis: A user-friendly guide for social scientists.* London: Routledge.

Dey, I. (1999). *Grounding grounded theory: Guidelines for qualitative inquiry.* New York: Academic Press.

Dialogue: On fieldwork in your own setting. (1991). In J. M. Morse (Ed.), *Qualitative nursing research: A contemporary dialogue* (p. 72). Newbury Park, CA: Sage.

Eisner, E. W., & Peshkin, A. (Eds.). (1998). *Qualitative inquiry in education: The continuing debate.* New York: Teachers College Press.

Ellis, C., & Bochner, A. P. (Eds.). (1996). *Composing ethnography: Alternative forms of qualitative writing.* Walnut Creek, CA: AltaMira.

Ellis, C., & Bochner, A. P. (2000). Autoethnography, personal narrative, reflexivity: Researcher as subject. In N. K. Denzin & Y. S. Lincoln (Eds.), *Handbook of qualitative research* (2nd ed., pp. 733–768). Thousand Oaks, CA: Sage.

Erlandson, D. A., Harris, E. L., Skipper, B. L., & Allen, S. D. (1993). *Doing naturalistic inquiry: A guide to methods.* Newbury Park, CA: Sage.

Ezzy, D., Liamputtong, P., & Hollis, D. B. (2005). *Qualitative research methods.* Oxford, UK: Oxford University Press.

Farber, N. G. (1990). Through the camera's lens: Video as a research tool. In I. Harel (Ed.), *Constructionist learning* (pp. 319–326). Cambridge: MIT Media Laboratory.

Fetterman, D. M. (1989). *Ethnography: Step by step.* Newbury Park, CA: Sage.

Fetterman, D. M. (1998). *Ethnography: Step by step* (2nd ed.). Thousand Oaks, CA: Sage.

Finch, J. (1984). "It's great to have someone to talk to": The ethics and politics of interviewing women. In C. Bell & H. Roberts (Eds.), *Social researching: Politics, problems, practice* (pp. 70–87). London: Routledge & Kegan Paul.

Finch, J. (1986). *Research and policy: The uses of qualitative research in social and educational research.* London: Falmer.

Finch, J. (1987). The vignette technique in survey research. *Sociology, 21,* 105–114.

Finch, J. (1989). *Family obligations and social change.* Cambridge, UK: Polity.

Fitzpatrick, J., Secrist, J., & Wright, D. J. (1998). *Secrets for a successful dissertation.* Thousand Oaks, CA: Sage.

Flick, U. (1998). *An introduction to qualitative research.* London: Sage.

Fontana, A., & Frey, J. H. (2000). The interview: From structured questions to negotiated text. In N. K. Denzin & Y. S. Lincoln (Eds.), *Handbook of qualitative research* (2nd ed., pp. 645–672). Thousand Oaks, CA: Sage.

Frank, A. W. (1991). *At the will of the body: Reflections on illness.* Boston: Houghton Mifflin.

Geertz, C. (1973). *The interpretation of cultures: Selected essays.* New York: Basic Books.

Germain, C. (1979). *The cancer unit: An ethnography.* Wakefield, MA: Nursing Resources.

Gibbs, G. (2002). *Qualitative data analysis: Explorations with NVivo.* London: Open University Press.

Gilgun, J. F., Daly, K., & Handel, G. (Eds.). (1992). *Qualitative methods in family research.* Newbury Park, CA: Sage.

Giorgi, A. (Ed.). (1985). *Phenomenology and psychological research.* Pittsburgh, PA: Duquesne University Press.

Giorgi, A. (1997). The theory, practice, and evaluation of the phenomenological methods as a qualitative research procedure. *Journal of Phenomenological Psychology, 28,* 235–281.

Glaser, B. G. (1978). *Theoretical sensitivity: Advances in the methodology of grounded theory.* Mill Valley, CA: Sociology Press.

Glaser, B. G. (1992). *Basics of grounded theory analysis: Emergence vs. forcing.* Mill Valley, CA: Sociology Press.

Glaser, B. G. (Ed.). (1993). *Examples of grounded theory: A reader.* Mill Valley, CA: Sociology Press.

Glaser, B. G. (Ed.). (1996). *Gerund grounded theory: The basic social process dissertation.* Mill Valley, CA: Sociology Press.

Glaser, B. G. (1998). *Doing grounded theory: Issues and discussions.* Mill Valley, CA: Sociology Press.

Glaser, B. G., & Strauss, A. L. (1967). *The discovery of grounded theory: Strategies for qualitative research.* Chicago: Aldine.

Glaser, B. G., & Strauss, A. L. (1968). *Time for dying.* Chicago: Aldine.

Glaser, B. G., & Strauss, A. L. (1971). *Status passage.* Chicago: Aldine.

Goffman, E. (1989). On fieldwork. *Journal of Contemporary Ethnography, 18,* 123–132.

Golding, C. (2002). *Grounded theory: A practical guide for management, business & market researchers.* Thousand Oaks, CA: Sage.

Goldman-Segall, R. (1998). *Points of viewing children's thinking: A digital ethnographer's journey.* Mahwah, NJ: Lawrence Erlbaum. (See also website at http://www.pointsofviewing.com.)

Grbich, C. (1999). *Qualitative research in health: An introduction.* Sydney: Allen and Unwin.

Greenwood, D. J., & Levin, M. (1998). *Introduction to action research: Social research for social change.* Thousand Oaks, CA: Sage.

Guba, E. G., & Lincoln, Y. S. (1989). *Fourth generation evaluation.* Newbury Park, CA: Sage.

Gubrium, J. F. (1975). *Living and dying at Murray Manor.* New York: St. Martin's.

Gubrium, J. F., & Holstein, J. A. (Eds.). (2002). *Handbook of interview research: Context and method.* Thousand Oaks, CA: Sage.

Hammersley, M., & Atkinson, P. (1983). *Ethnography: Principles in practice.* London: Tavistock.

Harel, I. (1991). The silent observer and holistic note taker: Using video for documenting a research project. In I. Harel & S. Papert (Eds.), *Constructionism* (pp. 449–464). Norwood, NJ: Ablex.

Hart, C. (1998). *Doing a literature review: Releasing the social science research imagination.* London: Sage.

Haug, F. (1987). *Female sexualization: A collective work of memory* (E. Carter, Trans.). London: Verso.

Holloway, I. (2005). *Qualitative research in health care.* Oxford, UK: Blackwell Science.

Holstein, J. A., & Gubrium, J. F. (2003). *Inside interviewing.* Thousand Oaks, CA: Sage.

Jorgensen, D. L. (1989). *Participant observation: A methodology for human studies.* Newbury Park, CA: Sage.

Karp, D. A. (1996). *Speaking of sadness: Depression, disconnection, and the meaning of illness.* New York: Oxford University Press.

Kelle, U. (Ed.). (1995). *Computer-aided qualitative data analysis.* London: Sage.

Kelpin, V. (1992). Birthing pain. In J. M. Morse (Ed.), *Qualitative health research* (pp. 93–103). Newbury Park, CA: Sage. (Original work published 1984)

King, N. M. P., Henderson, G. E., & Stein, J. (1999). *Beyond regulation: Ethics in human subjects research.* Chapel Hill: University of North Carolina Press.

Kools, S., McCarthy, M., Durham, R., & Robrecht, L. (1996). Dimensional analysis: Broadening the conception of grounded theory. *Qualitative Health Research, 6,* 312–330.

Krueger, R. (1994). *Focus groups: A practical guide for applied research.* Thousand Oaks, CA: Sage.

Krueger, R., & Casey, M. A. (2000). *Focus groups: A practical guide for applied research* (3rd ed.). Thousand Oaks, CA: Sage.

Kuzel, A. J., & Engel, J. D. (2001). Some pragmatic thoughts about evaluating qualitative research. In J. M. Morse, J. M. Swanson, & A. J. Kuzel (Eds.), *The nature of qualitative evidence* (pp. 114–139). Thousand Oaks, CA: Sage.

Kvale, S. (1989). *Issues of validity in qualitative research.* Lund, Sweden: Cartwell Bratt.

Kvale, S. (1995). The social construction of validity. *Qualitative Inquiry, 1,* 19–40.

Kvale, S. (1996). *InterViews: An introduction to qualitative research interviewing.* Thousand Oaks, CA: Sage.

Latimer, J. (Ed.). (2003). *Advanced qualitative research for nursing.* Oxford, UK: Blackwell.

LeCompte, M. D., Millroy, W. L., & Preissle, J. (Eds.). (1992). *The handbook of qualitative research in education.* San Diego, CA: Academic Press.

LeCompte, M. D., & Preissle, J. (1993). *Ethnography and qualitative design in educational research.* San Diego, CA: Academic Press.

LeCompte, M. D., & Schensul, J. J. (1999a). *Ethnographer's toolkit: Vol. 1. Designing and conducting ethnographic research* (J. J. Schensul & M. D. LeCompte, Series Eds.). Walnut Creek, CA: AltaMira.

LeCompte, M. D., & Schensul, J. J. (1999b). *Ethnographer's toolkit: Vol. 5. Analyzing and interpreting ethnographic data* (J. J. Schensul & M. D. LeCompte, Series Eds.). Walnut Creek, CA: AltaMira.

LeCompte, M. D., Schensul, J. J., Weeks, M. R., & Singer, M. (1999). *Ethnographer's toolkit: Vol. 6. Researcher roles and research partnerships* (J. J. Schensul & M. D. LeCompte, Series Eds.). Walnut Creek, CA: AltaMira.

Lee, R. M., & Fielding, N. G. (1996). Qualitative data analysis. Representations of a technology: A comment on Coffey, Holbrook and Atkinson. *Sociological Research Online, 1*(4). Retrieved May 8, 2001, from http://www.socresonline.org.uk/socresonline/1/4/lf.html

Leininger, M. (1994). Evaluation criteria and critique of qualitative research studies. In J. M. Morse (Ed.), *Critical issues in qualitative research methods* (pp. 95–115). Thousand Oaks, CA: Sage.

Lewins, A. F., & Silver, C. (in press). *Qualitative data analysis software: A step-by-step guide.* London: Sage.

Lincoln, Y. S. (1995). Emerging criteria for quality in qualitative and interpretive research. *Qualitative Inquiry, 1,* 275–289.

Lincoln, Y. S., & Guba, E.G. (1985). *Naturalistic inquiry.* Beverly Hills, CA: Sage.

Lipson, J. G. (1991). The use of self in ethnographic research. In J. M. Morse (Ed.), *Qualitative nursing research: A contemporary dialogue* (pp. 73–89). Newbury Park, CA: Sage.

Locke, L. F., Spirduso, W. W., & Silverman, S. J. (1993). *Proposals that work: A guide for planning dissertations and grant proposals* (3rd ed.). Newbury Park, CA: Sage.

Lofland, J., & Lofland, L. H. (1995). *Analyzing social settings: A guide to qualitative observation and analysis* (3rd ed.). Belmont, CA: Wadsworth.

Lomax, H., & Casey, N. (1998). Recording social life: Reflexivity and video methodology. *Social Research Online, 3*(2). Retrieved May 8, 2001, from http://www.socresonline.org.uk/socresonline/3/2/1.html

Lorencz, B. J. (1992). Becoming ordinary: Leaving the psychiatric hospital. In J. M. Morse (Ed.), *Qualitative health research* (pp. 259–318). Newbury Park, CA: Sage. (Original work published 1988)

Malacrida, C. (1998). *Mourning the dreams: How parents create meaning from miscarriage, stillbirth and early infant death.* Edmonton, AB: Qual Institute Press.

Mariampolski, H. (2001). *Qualitative market research: A comprehensive guide.* Thousand Oaks, CA: Sage.

Marshall, C., & Rossman, G. B. (1999). *Designing qualitative research* (3rd ed.). Thousand Oaks, CA: Sage.

Mason, J. (1996). *Qualitative researching.* London: Sage.

Mason, J. (2002). *Qualitative researching* (2nd ed.). London: Sage.

Mauthner, M., Birch, M., Jessop, J., & Mueller, T. (Eds.) (2002). *Ethics in qualitative research.* Thousand Oaks, CA. Sage.

Maxwell, J. A. (1992). Understanding validity in qualitative research. *Harvard Educational Review, 62,* 279–300.

Maxwell, J. A. (1996). *Qualitative research design: An interactive approach.* Thousand Oaks, CA: Sage.

Maxwell, J. A. (1998). Designing a qualitative study. In L. Bickman & D. J. Rog (Eds.), *Handbook of applied social research methods* (pp. 69–100). Thousand Oaks, CA: Sage.

May, K. (Ed.). (1996). Advances in grounded theory [Special issue]. *Qualitative Health Research, 6*(3).

May, T. (Ed.). (2002). *Qualitative research in action.* Thousand Oaks, CA: Sage.

Meadows, L., & Morse, J. M. (2001). Constructing evidence within the qualitative project. In J. M. Morse, J. M. Swanson, & A. J. Kuzel (Eds.), *The nature of qualitative evidence* (pp. 187–200). Thousand Oaks, CA: Sage.

Melia, K. M. (1997). Producing "plausible stories": Interviewing student nurses. In G. Miller & R. Dingwall (Eds.), *Context and method in qualitative research* (pp. 26–36). London: Sage.

Meloy, J. M. (1994). *Writing the qualitative dissertation: Understanding by doing.* Hillsdale, NJ: Lawrence Erlbaum.

Merriman, N. B. (1997). *Qualitative research and case study applications in education.* Toronto: John Wiley & Sons.

Miles, M. B., & Huberman, A. M. (1994). *Qualitative data analysis: An expanded sourcebook* (2nd ed.). Thousand Oaks, CA: Sage.

Miller, D. C. (1991). *Handbook of research design and social measurement* (5th ed.). Newbury Park, CA: Sage.

Miller, G., & Dingwall, R. (Eds.). (1997). *Context and method in qualitative research.* London: Sage.

Mithaug, D. E. (2000). *Learning to theorize.* Thousand Oaks, CA: Sage.

Morgan, D. (1993). *Successful focus groups: Advancing the state of the art.* Newbury Park, CA: Sage.

Morgan, D. (1997). *Focus groups as qualitative research.* Thousand Oaks, CA: Sage.

Morgan, D., & Krueger, R. (1997). *The focus group kit.* Thousand Oaks, CA: Sage.

Morse, J. M. (1989a). Cultural responses to parturition: Childbirth in Fiji. *Medical Anthropology, 12*(1), 35–44.

Morse, J. M. (1989b). Gift-giving in the patient-nurse relationship: Reciprocity for care? *Western Journal of Nursing Research, 13,* 597–615.

Morse, J. M. (1991). Approaches to qualitative and quantitative methodological triangulation. *Nursing Research, 40*(2), 120–123.

Morse, J. M. (1992a). Comfort: The refocusing of nursing care. *Clinical Nursing Research, 1,* 91–113.

Morse, J. M. (Ed.). (1992b). *Qualitative health research.* Newbury Park, CA: Sage.

Morse, J. M. (1994a). Designing funded qualitative research. In N. K. Denzin & Y. S. Lincoln (Eds.), *Handbook of qualitative research* (pp. 220–235). Thousand Oaks, CA: Sage.

Morse, J. M. (1994b). "Emerging from the data": The cognitive processes of analysis in qualitative inquiry. In J. M. Morse (Ed.), *Critical issues in qualitative research methods* (pp. 23–42). Thousand Oaks, CA: Sage.

Morse, J. M. (1996). Is qualitative research complete? *Qualitative Health Research, 6,* 3–5.

Morse, J. M. (1997). Considering theory derived from qualitative research. In J. M. Morse (Ed.), *Completing a qualitative project: Details and dialogue* (pp. 163–188). Thousand Oaks, CA: Sage.

Morse, J. M. (2001). Types of talk: Modes of responses and data-led analytic strategies. In P. L. Munhall (Ed.), *Nursing research: A qualitative perspective* (3rd ed., pp. 565–578). Boston: Jones & Bartlett.

Morse, J. M. (2002). Principles of mixed and multimethod design. In A. Tashakkori & C. Teddlie (Eds.), *Mixed methodology: Combining qualitative and quantitative approaches.* Thousand Oaks, CA: Sage.

Morse, J. M. (2003). A review committee's guide for evaluating qualitative proposals. *Qualitative Health Research, 13,* 833–851.

Morse, J. M. (2006). Strategies of intraproject sampling. In P. Munhall (Ed.), *Nursing research: A qualitative perspective* (4th ed., pp. 529—540). Boston: Jones & Bartlett.

Morse, J. M., & Bottorff, J. L. (1992). The emotional experience of breastfeeding. In J. M. Morse (Ed.), *Qualitative health research* (pp. 319–332). Newbury Park, CA: Sage. (Original work published 1988)

Morse, J. M., & Field, P. A. (1995). *Qualitative research methods for health professionals* (2nd ed.). Thousand Oaks, CA: Sage.

Morse, J. M., Hupcey, J. E., Penrod, J., Spiers, J. A., Pooler, F. C., & Mitcham, C. (2002). Issues and validity: Behavioral concepts, their derivation and interpretation. *International Journal of Qualitative Methods, 1*(4). http://www.ualberta.ca/~ijqm

Morse, J. M., Miles, M. W., Clark, D. A., & Doberneck, B. M. (1994). "Sensing" patient needs: Exploring concepts of nursing insight and receptivity in nursing assessment. *Scholarly Inquiry for Nursing Practice, 8,* 233–254.

Morse, J. M., Mitcham, C., & van der Steen, V. (1998). Compathy or physical empathy: Implications for the caregiver relationship. *Journal of Medical Humanities, 19*(1), 51–65.

Morse, J. M., & Proctor, A. (1998). Maintaining patient endurance: The comfort work of trauma nurses. *Clinical Nursing Research, 7,* 250–274.

Morse, J. M., Swanson, J. N., & Kuzel, H. A. (2001). *The nature of qualitative evidence.* Thousand Oaks, CA: Sage.

Morse, J. M., Wolfe, R., & Niehaus, L. (in press). Principles and procedures for maintaining validity for mixed method design. In Leslie Curry, Renée Shield, & Terrie Wetle (Eds.), *Qualitative Methods in Research and Public Health: Aging and Other Special Populations.* Washington, DC: GSA and APHA.

Moustakas, C. (1990). *Heuristic research: Design, methodology, and applications.* Newbury Park, CA: Sage.

Moustakas, C. (1994). *Phenomenological research methods.* Thousand Oaks, CA: Sage.

Muecke, M. A. (1994). On the evaluation of ethnographies. In J. M. Morse (Ed.), *Critical issues in qualitative research methods* (pp. 187–209). Thousand Oaks, CA: Sage.

Munhall, P. L. (1994). *Revisioning phenomenology* (Publication No. 41–2545). New York: National League for Nursing Press.

Munhall, P. L. (2001). *Nursing research: A qualitative perspective* (3rd ed.). Boston: Jones & Bartlett.

National Institutes of Health, Office for Human Research Protections. (2001). *Protection of human subjects* (Title 45, Code of Federal Regulations, Part 46). Retrieved November 18, 2001, from http://ohsr.od.nih.gov/mpa/45cfr46.php3

Nunnally, J. C. (1978). *Psychometric theory* (2nd ed.). New York: McGraw-Hill.

Oakley, A. (1981). Interviewing women: A contradiction in terms? In H. Roberts (Ed.), *Doing feminist research* (pp. 30—61). London: Routledge.

Olesen, V. L. (2000). Feminisms and qualitative research at and into the millennium. In N. K. Denzin & Y. S. Lincoln (Eds.), *Handbook of qualitative research* (2nd ed., pp. 215–255). Thousand Oaks, CA: Sage.

Patton, M. Q. (2002). *Qualitative research & evaluation methods* (3rd ed.). Thousand Oaks, CA: Sage.

Penrod, J., & Morse, J. M. (1997). Strategies for assessing and fostering hope: The *Hope Assessment Guide. Oncology Nurses Forum, 24*(6), 1055–1063.

Piantanida, M., & Garman, N. B. (1999). *The qualitative dissertation: A guide for students and faculty.* Thousand Oaks, CA: Sage.

Potter, J., & Wetherell, M. (1994). Analyzing discourse. In A. Bryman & R. G. Burgess (Eds.), *Analyzing qualitative data* (pp. 47–67). London: Routledge.

Prasad, P. (2005). *Crafting qualitative research: Working in the post positivist tradition.* Armonk, NY: M. E. Sharpe.

Proctor, A., Morse, J. M., & Khonsari, E. S. (1996). Sounds of comfort in the trauma centre: How nurses talk to patients in pain. *Social Science & Medicine, 42,* 1669–1680.

Ray, M. A. (1994). The richness of phenomenology: Philosophic, theoretic, and methodologic concerns. In J. M. Morse (Ed.), *Critical issues in qualitative research methods* (pp. 117–133). Thousand Oaks, CA: Sage.

Reason, P. (Ed.). (1988). *Human inquiry in action: Developments in new paradigm research.* London: Sage.

Reason, P., & Bradbury, H. (2001). *Handbook of action research.* London: Sage.

Richards, L. (1990). *Nobody's home: Dreams and realities in a new suburb.* Melbourne: Oxford University Press.

Richards, L. (1998). Closeness to data: The changing goals of qualitative data handling. *Qualitative Health Research, 8,* 319–328.

Richards, L. (1999a). Data alive! The thinking behind NVivo. *Qualitative Health Research, 9,* 412–428.

Richards, L. (1999b). Qualitative teamwork: Making it work. *Qualitative Health Research, 9,* 7–10.

Richards, L. (2000, September). *Pattern analysis and why it isn't grounded theory.* Paper presented at the Conference on Strategies for Qualitative Research Using QSR Software, University of London.

Richards, L. (2005). *Handling qualitative data: A practical guide.* London: Sage.

Richards, L., & Richards, T. J. (1994). From filing cabinet to computer. In A. Bryman & R. G. Burgess (Eds.), *Analyzing qualitative data* (pp. 146–172). London: Routledge.

Richards, L., Seibold, C., & Davis, N. (1997). *Intermission: Women's experiences of midlife and menopause.* Melbourne: Oxford University Press.

Richards, T. J., & Richards, L. (1994). Using computers in qualitative research. In N. K. Denzin & Y. S. Lincoln (Eds.), *Handbook of qualitative research* (pp. 445–462). Thousand Oaks, CA: Sage.

Richards, T. J., & Richards, L. (1995). Using hierarchical categories in qualitative data analysis. In U. Kelle (Ed.), *Computer-aided qualitative data analysis: Theory, methods, and practice* (pp. 62–68). London: Sage.

Richardson, L. (1990). *Writing strategies: Reaching diverse audiences.* Newbury Park, CA: Sage.

Richardson, L. (1994). Writing: A method of inquiry. In N. K. Denzin & Y. S. Lincoln (Eds.), *Handbook of qualitative research* (pp. 516–529). Thousand Oaks, CA: Sage.

Riessman, C. K. (1993). *Narrative analysis.* Newbury Park, CA: Sage.

Riessman, C. K. (Ed.). (1994). *Qualitative studies in social work research.* Thousand Oaks, CA: Sage.

Ritchie, J., & Lewis, J. (Eds.) (2004). *Qualitative research practice. A guide for social science students and researchers.* Thousand Oaks, CA: Sage.

Rossman, G. B., & Rallis, S. (1998). *Learning in the field: An introduction to qualitative research.* Thousand Oaks, CA: Sage.

Rossman, M. H. (1995). *Negotiating graduate school: A guide for graduate students.* Thousand Oaks, CA: Sage.

Rothe, J. P. (2000). *Undertaking qualitative research: Concepts and cases in injury, health and social life.* Edmonton: University of Alberta Press.

Rubin, H., & Rubin, I. (1995). *Qualitative interviewing: The art of hearing data.* Thousand Oaks, CA: Sage.

Rudestam, K. E., & Newton, R. B. R. (1992). *Surviving your dissertation: A comprehensive guide to content and process.* Newbury Park, CA: Sage.

Sandelowski, M. (1993). Rigor or rigor mortis: The problem of rigor in qualitative inquiry. *Advances in Nursing Science, 16*(2), 1–8.

Sapsford, R., & Jupp, V. (Eds.). (1996). *Data collection and analysis.* London: Sage.

Schatzman, L. (1991). Dimensional analysis: Notes on the alternative approach to the grounding of theory in qualitative research. In D. R. Maines (Ed.), *Social organization and social process* (pp. 303–314). New York: Aldine.

Schensul, J. J., & LeCompte, M. D. (Series Eds.). (1999). *Ethnographer's toolkit* (7 vols.). Walnut Creek, CA: AltaMira.

Schensul, J. J., LeCompte, M. D., Hess, G. A., Nastasi, B. K., Berg, M. J., Williamson, L., Brecher, J., & Glasser, R. (1999). *Ethnographer's toolkit: Vol. 7. Using ethnographic data* (J. J. Schensul & M. D. LeCompte, Series Eds.). Walnut Creek, CA: AltaMira.

Schensul, J. J., LeCompte, M. D., Nastasi, B. K., & Schensul, S. L. (1999). *Ethnographer's toolkit: Vol. 3. Enhanced ethnographic methods* (J. J. Schensul & M. D. LeCompte, Series Eds.). Walnut Creek, CA: AltaMira.

Schensul, J. J., LeCompte, M. D., Trotter, R. T., II, Cromley, E. K., & Singer, M. (1999). *Ethnographer's toolkit: Vol. 4. Mapping social networks, spatial data, and hidden populations* (J. J. Schensul & M. D. LeCompte, Series Eds.). Walnut Creek, CA: AltaMira.

Schensul, S. L., Schensul, J. J., & LeCompte, M. D. (1999). *Ethnographer's toolkit: Vol. 2. Essential ethnographic methods* (J. J. Schensul & M. D. LeCompte, Series Eds.). Walnut Creek, CA: AltaMira.

Schreiber, R. S., & Stern, P. N. (Eds.). (2001). *Using grounded theory in nursing.* New York: Springer.

Schwandt, T. A. (1997). *Qualitative inquiry: A dictionary of terms.* Thousand Oaks, CA: Sage.

Seale, C. (1999). *Quality of qualitative research.* London: Sage.

Seale, C., Gobo, G., Gubrium, J., & Silverman, D. (2004). *Qualitative research practice.* Thousand Oaks, CA: Sage.

Shaw, I. S., & Gould, N. (2001). *Qualitative research and social work.* Thousand Oaks, CA: Sage.

Sieber, J. E. (1992). *Planning ethically responsible research: A guide for students and internal review boards.* Newbury Park, CA: Sage.

Silverman, D. (Ed.). (1997). *Qualitative research: Theory, method and practice.* London: Sage.

Silverman, D. (2005). *Doing qualitative research* (2nd ed.). London: Sage.

Smith, S. J. (1992). Operating on a child's heart: A pedagogical view of hospitalization. In J. M. Morse (Ed.), *Qualitative health research* (pp. 104–122). Newbury Park, CA: Sage. (Original work published 1989)

Sparkes, A. C. (2001). Qualitative health researchers will agree about validity. *Qualitative Health Research, 11,* 538–552.

Spiegelberg, H. (1975). *Doing phenomenology: Essays on and in phenomenology.* The Hague, Netherlands: Martinus Nijhoff.

Spradley, J. P. (1970). *You owe yourself a drunk: An ethnography of urban nomads.* Boston: Little, Brown.

Spradley, J. P. (1979). *The ethnographic interview.* New York: Holt, Rinehart & Winston.

Spradley, J. P. (1980). *Participant observation.* New York: Holt, Rinehart & Winston.

Stern, P. N. (1994). Eroding grounded theory. In J. M. Morse (Ed.), *Critical issues in qualitative research methods* (pp. 212–223). Thousand Oaks, CA: Sage.

Stern, P. N., & Kerry, J. (1996). Restructuring life after home loss by fire. *Image: Journal of Nursing Scholarship, 28,* 9–14.

Stewart, D., & Shamdasani, P. (1990). *Focus groups: Theory and practice.* Newbury Park, CA: Sage.

Strauss, A. L. (1987). *Qualitative analysis for social scientists.* New York: Cambridge University Press.

Strauss, A. L. (1995). Notes on the nature and development of general theories. *Qualitative Inquiry, 1,* 7–18.

Strauss, A. L., & Corbin, J. (1990). *Basics of qualitative research: Grounded theory procedures and techniques.* Newbury Park, CA: Sage.

Strauss, A. L., & Corbin, J. (1994). Grounded theory methodology: An overview. In N. K. Denzin & Y. S. Lincoln (Eds.), *Handbook of qualitative research* (pp. 273–285). Thousand Oaks, CA: Sage.

Strauss, A. L., & Corbin, J. (Eds.). (1997). *Grounded theory in practice.* Thousand Oaks, CA: Sage.

Strauss, A. L., & Corbin, J. (1998). *Basics of qualitative research: Techniques and procedures for developing grounded theory* (2nd ed.). Thousand Oaks, CA: Sage.

Tedlock, B. (2000). Ethnography and ethnographic representation. In N. K. Denzin & Y. S. Lincoln (Eds.), *Handbook of qualitative research* (2nd ed., pp. 455–486). Thousand Oaks, CA: Sage.

Tesch, R. (1990). *Qualitative research: Analysis types and software tools.* London: Falmer.

Thomas, J. (1993). *Doing critical ethnography.* Newbury Park, CA: Sage.

Thompson, P. R. (1998). Sharing and reshaping life stories: Problems and potential in archiving research narratives. In M. Chamberlain & P. R. Thompson (Eds.), *Narrative and genre* (pp. 167–181). London: Routledge.

Thompson, P. R., Itzen, C., & Abendstern, M. (1991). *"I don't feel old": Understanding the experience of later life.* Oxford, UK: Oxford University Press.

Thorne, S. (1997). The art (and science) of critiquing qualitative research. In J. M. Morse (Ed.), *Completing a qualitative project: Details and dialogue* (pp. 117–132). Thousand Oaks, CA: Sage.

Tierney, W. G. (2000). Undaunted courage: Life history and the postmodern challenge. In N. K. Denzin & Y. S. Lincoln (Eds.), *Handbook of qualitative research* (2nd ed., pp. 537–553). Thousand Oaks, CA: Sage.

Turner, B. A. (1981). Some practical aspects of qualitative data analysis. *Quality and Quantity, 15,* 225–247.

Turner, B. A. (1994). Patterns of crisis behaviour: A qualitative inquiry. In A. Bryman & R. G. Burgess (Eds.), *Analyzing qualitative data* (pp. 195–216). London: Routledge.

Ulin, P. R., Robinson, E. T., & Tolley, E. E. (2005). *Qualitative methods in public health: A field guide for applied research.* San Francisco, CA: Jossey-Bass.

Van den Hoonard, W. C. (1999). *Working with sensitizing concepts.* Thousand Oaks, CA: Sage.

Van Maanen, J. (1988). *Tales of the field: On writing ethnography.* Chicago: University of Chicago Press.

Van Maanen, J. (Ed.). (1995). *Representation in ethnography.* Thousand Oaks, CA: Sage.

Van Maanen, J., Dabbs, J. M., & Faulkner, R. R. (1982). *Varieties of qualitative research.* Beverly Hills, CA: Sage.

van Manen, M. (1990). *Researching lived experience: Human science for an action sensitive pedagogy.* London, ON: Althouse.

van Manen, M. (1991). *The tact of teaching: The meaning of pedagogical thoughtfulness.* London, ON: Althouse.

van Manen, M. (1997). From meaning to method. *Qualitative Health Research, 7,* 345–369.

van Manen, M. (2006). http://www.phenomenologyonline.com/inquiry/2.html [downloaded April 3, 2006]

van Manen, M. (Ed.). (n.d.). *Textorium.* Retrieved May 8, 2001, from http://www .ualberta.ca/~vanmanen/textorium.html

Wax, R. H. (1971). *Doing fieldwork: Warnings and advice.* Chicago: University of Chicago Press.

Weitzman, E., & Miles, M. B. (1995). *Computer programs for qualitative data analysis.* Thousand Oaks, CA: Sage.

Werner, O., & Schoepfle, G. M. (1987a). *Systematic fieldwork: Vol. 1. Foundations of ethnography and interviewing.* Beverly Hills, CA: Sage.

Werner, O., & Schoepfle, G. M. (1987b). *Systematic fieldwork: Vol. 2. Ethnographic analysis and data management.* Beverly Hills, CA: Sage.

Westphal, L. M. (2000). Increasing the trustworthiness of research results: The role of computers in qualitative text analysis. In D. N. Bengston (Ed.), *Applications of computer-aided text analysis in natural resources* (General Technical Report No. NC-211). St. Paul, MN: USDA Forest Service, North Central Research Station.

Whittemore, R., Chase, S. K., & Mandle, C. L. (2001). Validity in qualitative research. *Qualitative Health Research, 11,* 522–537.

Whyte, W. F. (Ed.). (1991). *Participatory action research.* London: Sage.

Wilson, H. S., & Hutchinson, S. A. (1991). Combining Heideggerian hermeneutics and grounded theory methods to study caregiver decision-making with elderly Alzheimer's patients. *Qualitative Health Research, 1,* 263–276.

Wilson, H. S., & Hutchinson, S. A. (1996). Methodologic mistakes in grounded theory. *Nursing Research, 45*(2), 122–124.

Wolcott, H. F. (1994). *Transforming qualitative data: Description, analysis, and interpretation.* Thousand Oaks, CA: Sage.

Wolcott, H. F. (1995). *The art of fieldwork.* Thousand Oaks, CA: Sage.

Wolcott, H. F. (1999). *Ethnography: A way of seeing.* Walnut Creek, CA: AltaMira.

Index

Abstract thinking, 153
Abstracting
  additional resources for, 167–168
  categorizing step of, 155–157
  conceptualizing step of, 157
  managing, 161–164
  overview of, 153–154
  process of doing, 158–161
  using software for managing ideas
    of, 164–166
  terminology associated with, 154
Abstraction management
  documenting ideas, 162
  growing ideas, 162–163
  importance of, 161–162
  index systems used for, 163
  models and diagrams used for, 164
Abstraction process
  approaches taken by three major
    methods, 159*t*, 160–161
  theory emergence versus theory
    construction using, 161
  when it happens, 158, 160
Action research (AR), 59
Actors, 126
Altheide, D. L., 190
Analysis. *See* Data analysis
Analysis techniques. *See* Techniques
Analytic coding, 141–143
Anonymity issue, 235–236
Arenas maps/social networks, 65
Armchair walkthrough, 26–27,
  36–37, 244–245
Article writing

issues to consider in, 219–220
process and steps of, 220–222
Qualitative Health Research
  review criteria for
  209*b*–210*b*, 220–222
Assessment of saturation, 196
At the Will of the Body (Frank), 127
Atkinson, P., 133
Audit (or log) trail, 199–200
Autoethnography, 58
Awareness of self, 56
Axial coding, 142

Bad data/good data, 109–110
Basic social process (BSP),
  60, 62, 182
Basic social psychological process
  (BSPP), 60
Bates, G., 59
Bazeley, P., 98, 133
Becker, H. S., 207
Being in the world
  assumption, 50
Benner, P., 52
Bibliography, 216–217
Blumer, H., 59
Bochner, A. P., 58, 127
"Body Image" project, 109
Bottorff, J. L., 60
Boyd, C. O., 51, 172, 230
Boyle, J. S., 53
Bracketing, 170–172, 191–192
Bryman, A., 98
Budget issues, 232, 234

About the Authors

Lyn Richards (B.A. Hon., political science; M.A., sociology) is a qualitative research writer and consultant and Adjunct Professor at the International Institute of Qualitative Methodology in Canada. She is also a founder and Director of QSR International, Melbourne. As a family sociologist, she has published four books and many papers on Australian families and women's roles. As a methodologist, she taught graduate and undergraduate qualitative research at La Trobe University and went on to write for and teach the teachers. Her tenth book is *Handling Qualitative Data* (2005). In university research with Tom Richards, she developed the NUD*IST software. In interaction with the researchers using it, and later the development teams at QSR, she worked on the design of the subsequent versions (to N6) and the new-generation program, NVivo, as a principal member of the QSR software-development teams and author of the software's documentation. She has been an invited speaker at all of the conferences on qualitative computing and is a leading teacher and trainer internationally in qualitative computing and the handling of qualitative data. Richards has taught qualitative methods and qualitative software to some 4,000 researchers in 15 countries, and learned from them all. She can be contacted via her website, www.lynrichards.org.

Janice M. Morse (R.N.; Ph.D., anthropology; Ph.D., nursing; D.Nurs. Hon.; FAAN) is Scientific Director, International Institute of Qualitative Methodology; Professor, Faculty of Nursing, University of Alberta; and Adjunct Professor, School of Nursing, Pennsylvania State University. She has an interest in developing qualitative methods and has published more than 300 articles and 14 books on clinical nursing research and research methods. She had edited and authored several books, most recently *The Nature of Qualitative Evidence* (with J. M. Swanson and A. J. Kuzel, 2001). She is the editor of *Qualitative Health Research,* an interdisciplinary journal publishing on qualitative methods and research, and of the

International Journal of Qualitative Methods. She was the 1997 Sigma Theta Tau Episteme Laureate, and in 1999 she received an honorary doctorate from the University of Newcastle, Australia, for her contribution to nursing knowledge. She is currently funded by CIHR to conduct a qualitative study on the risks inherent in qualitative interviews.